Precision Agriculture

Precision Agriculture
Evolution, Insights and Emerging Trends

Edited by

Qamar Zaman

Department of Engineering, Faculty of Agriculture, Dalhousie University, Halifax, NS, Canada
University of Arid Agriculture Rawalpindi, Rawalpindi, Punjab, Pakistan

Academic Press is an imprint of Elsevier
125 London Wall, London EC2Y 5AS, United Kingdom
525 B Street, Suite 1650, San Diego, CA 92101, United States
50 Hampshire Street, 5th Floor, Cambridge, MA 02139, United States
The Boulevard, Langford Lane, Kidlington, Oxford OX5 1GB, United Kingdom

Copyright © 2023 Elsevier Inc. All rights reserved.

No part of this publication may be reproduced or transmitted in any form or by any means, electronic or mechanical, including photocopying, recording, or any information storage and retrieval system, without permission in writing from the publisher. Details on how to seek permission, further information about the Publisher's permissions policies and our arrangements with organizations such as the Copyright Clearance Center and the Copyright Licensing Agency, can be found at our website: www.elsevier.com/permissions.

This book and the individual contributions contained in it are protected under copyright by the Publisher (other than as may be noted herein).

Notices
Knowledge and best practice in this field are constantly changing. As new research and experience broaden our understanding, changes in research methods, professional practices, or medical treatment may become necessary.

Practitioners and researchers must always rely on their own experience and knowledge in evaluating and using any information, methods, compounds, or experiments described herein. In using such information or methods they should be mindful of their own safety and the safety of others, including parties for whom they have a professional responsibility.

To the fullest extent of the law, neither the Publisher nor the authors, contributors, or editors, assume any liability for any injury and/or damage to persons or property as a matter of products liability, negligence or otherwise, or from any use or operation of any methods, products, instructions, or ideas contained in the material herein.

ISBN: 978-0-443-18953-1

For information on all Academic Press publications visit our website at https://www.elsevier.com/books-and-journals

Publisher: Nikki Levy
Acquisitions Editor: Nancy Maragioglio
Editorial Project Manager: Dan Egan
Production Project Manager: Paul Prasad Chandramohan
Cover Designer: Mark Rogers

Typeset by TNQ Technologies

I dedicate this book to my daughter Zarnab Qamar

Contents

List of contributors xvii
Foreword xxi
Preface xxiii
List of abbreviations xxv

CHAPTER 1 Precision agriculture technology: a pathway toward sustainable agriculture 1
Qamar U. Zaman

    1.1 Introduction 1
        1.1.1 Background 1
        1.1.2 Footprints of agricultural practices and products 2
        1.1.3 Need for agricultural sustainability 2
    1.2 Climate change, food security, and precision agriculture nexus .. 3
        1.2.1 Threats of climate change 3
        1.2.2 Food security challenges 3
        1.2.3 Solutions embedded in precision agriculture adoption 3
    1.3 Agricultural resources management 4
        1.3.1 Soil management 5
        1.3.2 Nutrient management 5
        1.3.3 Water management 5
        1.3.4 Supplemental light and CO_2 6
    1.4 Precision agriculture tools, techniques, and technologies 7
        1.4.1 Precision agriculture 8
        1.4.2 Precision agriculture tools 9
        1.4.3 Precision agriculture techniques 10
        1.4.4 Precision agriculture technologies 12
    1.5 This book 14
    1.6 Conclusions 15
    References 16

CHAPTER 2 Soil spatial variability and its management with precision agriculture 19
Humna Khan, Travis J. Esau, Aitazaz A. Farooque, Qamar U. Zaman, Farhat Abbas and Arnold W. Schumann

    2.1 Introduction 19
        2.1.1 Concepts of variability and precision agriculture 19
    2.2 Soil and crop variability 20
        2.2.1 Factors affecting soil and crop variability 21
        2.2.2 Monitoring variability in agricultural fields 21

2.3 Precision agriculture 22
2.3.1 The need for precision agriculture 23
2.3.2 Applications of precision agriculture technology 23
2.3.3 Data analysis and delineation of management zones 27
2.3.4 Variable-rate application technology 28
2.3.5 Precision irrigation management to manage spatial variability in soil water content 29
2.3.6 Precision agriculture using machine learning 30
2.3.7 Role of precision agriculture in climate change 31
2.3.8 Sustainable agriculture: the ultimate goal 32
2.4 Conclusion 32
References 33

CHAPTER 3 Geospatial technologies for the management of pest and disease in crops 37
Manjeet Singh, Aseem Vermaa and Vijay Kumar

3.1 Introduction and objectives 37
3.2 Brief review of literature 38
3.2.1 Case studies 40
3.3 Precision in pest management 41
3.4 Overview of scouting 43
3.4.1 Pest identification 43
3.4.2 Insect/pest assessment and control 44
3.5 Integrated pest management (IPM) 44
3.6 Sensing systems for detection and monitoring of pests and disease 45
3.7 Spatially variable rate technology for spraying 48
3.8 Opportunities and challenges 48
References 51

CHAPTER 4 Application of unmanned aerial vehicles in precision agriculture 55
Muhammad Naveed Tahir, Yubin Lan, Yali Zhang, Huang Wenjiang, Yingkuan Wang and Syed Muhammad Zaigham Abbas Naqvi

4.1 Introduction 55
4.2 Types of UAVs 56
4.2.1 Multirotor UAVs 56
4.2.2 Fixed-wing UAVs 56
4.2.3 Single-rotor UAVs 56
4.2.4 Fixed-wing multirotor hybrid UAVs 57
4.2.5 A hybrid fixed-wing multirotor UAV 57

4.3 Recent advances in the UAV technology 58
4.4 Literature review..58
    4.4.1 Use of UAVs for crop health monitoring 58
    4.4.2 UAVs for yield estimation.. 60
    4.4.3 Role of UAVs in pest detection................................... 62
    4.4.4 Role of UAVs for disease detection 63
4.5 The UAV-based precision agriculture chemical spraying application 63
4.6 Artificial intelligence for UAVs..65
4.7 Challenges and limitations of UAV technology adoption 66
4.8 Conclusion..67
References..67

CHAPTER 5 Applications of geospatial technologies for precision agriculture...71
Mobushir R. Khan, Richard A. Crabbe,
Naeem A. Malik and Lachlan O'Meara

5.1 Introduction..71
5.2 Management philosophy of PA 72
5.3 Enabling technologies of PA 72
    5.3.1 Geographic Information Systems 73
    5.3.2 Remote sensing (RS) 73
    5.3.3 Big data analytics... 75
5.4 State-of-the-art conceptual system....................................76
    5.4.1 Soil management system 76
    5.4.2 Auto steering system... 77
    5.4.3 Precise seeding system.. 77
    5.4.4 Variable rate fertilizer application 77
    5.4.5 Variable rate irrigation 77
    5.4.6 Variable rate sprayer.. 77
    5.4.7 Crop stress monitoring system.................................... 78
    5.4.8 Yield monitoring system... 78
5.5 Perceptions of digital technologies 80
5.6 Conclusions..81
References..82

CHAPTER 6 Precision irrigation: challenges and opportunities 85
Muhammad Naveed Anjum, Muhammad Jehanzeb Masud
Cheema, Fiaz Hussain and Ray-Shyan Wu

6.1 Introduction..85

6.2 Irrigation application methods 87
      6.2.1 Surface irrigation methods 89
      6.2.2 Pressurized irrigation method—sprinkler irrigation........ 89
      6.2.3 Micro-irrigation systems.. 91
6.3 Literature review..93
      6.3.1 Conceptualization of precision irrigation 93
      6.3.2 Benefits of precision irrigation.................................. 94
      6.3.3 Main components of precision irrigation 94
      6.3.4 Data collection... 94
      6.3.5 Data interpretation/analysis....................................... 95
      6.3.6 Control.. 95
      6.3.7 Assessment 96
6.4 Existing tools and technologies for precision irrigation 96
6.5 Data acquisition tools...97
      6.5.1 Weather data acquisition... 97
      6.5.2 Plant-based data sensors 97
      6.5.3 Soil moisture detections 98
      6.5.4 Remotely sensed data for irrigation scheduling............ 99
6.6 Challenges and opportunities..99
6.7 Conclusions..100
References..100

CHAPTER 7 Variable rate technologies: development, adaptation, and opportunities in agriculture.. 103
Shoaib Rashid Saleem, Qamar U. Zaman, Arnold W. Schumann and Syed Muhammad Zaigham Abbas Naqvi

7.1 Introduction...103
7.2 Variable rate technologies development............................... 105
      7.2.1 Cropping systems...105
      7.2.2 Sensing technologies...107
7.3 Adaption of precision agriculture techniques....................... 110
      7.3.1 User-friendliness of VRTs...111
      7.3.2 Artificial intelligence and variable rate technologies 112
7.4 Challenges for farmers and researchers 113
7.5 Future of variable rate technologies.................................... 115
7.6 Conclusions.. 116
References..117

CHAPTER 8 Yield monitoring and mechanical harvesting of wild blueberries to improve farm profitability.........123
Karen Esau, Qamar U. Zaman, Aitazaz A. Farooque, Travis J. Esau, Arnold W. Schumann and Farhat Abbas

    8.1 Introduction..123
        8.1.1 Cultivation and arvesting of wild blueberries..............124
    8.2 Factors affecting ripening of wild blueberry.......................126
        8.2.1 Meteorological factors..126
    8.3 Precision harvesting technologies.....................................127
        8.3.1 Challenge with hand-raking....................................128
        8.3.2 History of mechanical harvesters..............................128
        8.3.3 Working principle of wild blueberry mechanical harvester..129
    8.4 Factors affecting the mechanical harvesting 130
        8.4.1 Weather-related factors...131
        8.4.2 Human-induced factors—operator's skills..................133
        8.4.3 Field topography and vegetative conditions................134
        8.4.4 Mechanical factors 134
    8.5 Fruit losses during harvesting 134
        8.5.1 Prediction of yield losses.......................................135
        8.5.2 Yield mapping and mitigation of fruit losses...............136
    8.6 Conclusions...137
    References..137

CHAPTER 9 Artificial intelligence and deep learning applications for agriculture....................................141
Travis J. Esau, Patrick J. Hennessy, Craig B. MacEachern, Aitazaz A. Farooque, Qamar U. Zaman and Arnold W. Schumann

    9.1 Artificial intelligence, machine learning, and deep neural networks..141
    9.2 Machine learning approaches 142
        9.2.1 Supervised machine learning...................................142
        9.2.2 Unsupervised machine learning 144
    9.3 Deep neural network approaches......................................144
        9.3.1 Artificial neural networks 144
        9.3.2 Convolutional Neural Networks...............................145
    9.4 Machine learning applications in agriculture 145
        9.4.1 Crop management 146
        9.4.2 Livestock management..147

	9.4.3 Soil management...148
	9.4.4 Water management...148
9.5	Case study: deep neural networks for ripeness detection in wild blueberry..149
9.6	Applications in precision agriculture...151
	9.6.1 Autoguidance...151
	9.6.2 Uncrewed aerial vehicles..152
	9.6.3 Ground robots..153
	9.6.4 Precision irrigation...154
	9.6.5 Variable rate and spot specific agrochemical application..155
9.7	Conclusions and future implications..158
References..158

CHAPTER 10 Artificial neural modeling for precision agricultural water management practices 169

Hassan Afzaal, Aitazaz A. Farooque, Travis J. Esau, Arnold W. Schumann, Qamar U. Zaman, Farhat Abbas and Melanie Bos

10.1	Introduction..169
10.2	Common machine learning problems..171
10.3	Common deep learning frameworks..171
	10.3.1 TensorFlow..172
	10.3.2 Keras..172
	10.3.3 PyTorch..172
	10.3.4 MatLab...173
10.4	Popular machine learning architectures..173
	10.4.1 Multiplayer perceptron...173
	10.4.2 Recurrent neural networks...174
	10.4.3 Convolutional neural networks..174
	10.4.4 Gradient boosting...175
10.5	Machine learning model development steps...................................175
	10.5.1 Data preprocessing...175
	10.5.2 Feature importance..175
	10.5.3 Data split and model development..................................176
	10.5.4 Hyperparameter tuning..177
10.6	Application of machine learning in different hydrological fields..178
	10.6.1 Rainfall-runoff modeling...178
	10.6.2 Groundwater level modeling...178

    10.6.3 Evapotranspiration modeling 180
    10.6.4 Water resource management 181
  10.7 Ethical concerns and challenges in machine learning 182
  10.8 A case study 182
    10.8.1 Machine learning—based groundwater estimation 182
  10.9 Conclusion 183
  References 183

CHAPTER 11 Precision agriculture: making agriculture sustainable 187
Aneela Afzal and Mark Bell

  11.1 Introduction 187
    11.1.1 Evolution of precision agriculture 188
    11.1.2 Structure of precision agriculture 189
  11.2 Methodology 189
    11.2.1 PA and sustainability objectives of agricultural enterprises 190
    11.2.2 Technology adoption in agriculture 190
    11.2.3 Concept to practice 191
  11.3 Case study of PA adoption in Punjab, Pakistan 191
    11.3.1 Water-smart 193
    11.3.2 Soil and nutrient smart 193
  11.4 Evaluation of factors that hamper technology adoption in agriculture 194
  11.5 Financial and commercial sustainability of precision agriculture 196
    11.5.1 Evaluation of major cropwise yield gains, cost savings through precision agriculture 196
    11.5.2 Evaluation of financial viability of precision agriculture by employing popular capital budgeting techniques 197
    11.5.3 Addressing commercial barriers in adoption of precision agriculture 198
    11.5.4 Environmental impacts of conventional agriculture practices 198
    11.5.5 Ecological impacts of conventional agriculture practices 198
    11.5.6 Environmental impacts of precision agriculture 201
    11.5.7 Ecological impacts of precision agriculture 201

　　　　　　11.5.8 The verdict: environmental and ecological impacts of
　　　　　　　　　precision agriculture compared with conventional
　　　　　　　　　agriculture 203
　　11.6 Social and economic impact of precision agriculture 204
　　　　　　11.6.1 Precision agriculture's impact on farm household
　　　　　　　　　incomes 204
　　　　　　11.6.2 Economic dynamics of adoption of precision
　　　　　　　　　agriculture 204
　　　　　　11.6.3 Social and gender dynamics of precision
　　　　　　　　　agriculture 205
　　11.7 Conclusion 206
　　References 207

CHAPTER 12　Environment: role of precision agriculture technologies 211

Shoaib Rashid Saleem, Jana Levison and Zainab Haroon

　　12.1 Introduction 211
　　12.2 Environmental impacts of precision agriculture
　　　　　techniques 213
　　　　　　12.2.1 Surface water 214
　　　　　　12.2.2 Groundwater 215
　　　　　　12.2.3 Air 218
　　12.3 Potential climate change impacts 219
　　　　　　12.3.1 Extreme events and role of precision agriculture 220
　　　　　　12.3.2 Nutrient management based on spatiotemporal
　　　　　　　　　climate 221
　　12.4 Challenges for farmers and researchers 221
　　12.5 Conclusions 222
　　References 222

CHAPTER 13　Precision agriculture technologies: present adoption and future strategies 231

Muhammad Jehanzeb Masud Cheema, Tahir Iqbal, Andre Daccache, Saddam Hussain and Muhammad Awais

　　13.1 Introduction 231
　　　　　　13.1.1 Concept of precision agriculture 233
　　　　　　13.1.2 Developments in agricultural technologies 234
　　13.2 The transformation from mechanized to precision agriculture
　　　　　technologies 235

       13.2.1 Autosteering technology 236
       13.2.2 Land leveling to GNSS-based land leveling system .. 237
       13.2.3 Drills to planters and transplanters 237
       13.2.4 Mechanical sprayers to site-specific sprayers 237
       13.2.5 Ground sprayer to aerial spraying system 238
       13.2.6 Traditional to smart irrigation 238
       13.2.7 Mechanical harvesters to GPS-based smart
             harvesting machinery 239
       13.2.8 Remote sensing and GIS/GPS in agriculture 240
13.3 Adoptability of precision agriculture technologies in
the field 240
       13.3.1 Adoptability in the developed world 241
       13.3.2 Adoptability in the developing world 242
13.4 Case studies from developed and developing countries 243
13.5 Yield improvement and farm profitability through the
adoption of PA technologies 244
13.6 Problem/issues in adoptability of precision technologies 245
13.7 Future prospects of technology adoption 246
13.8 Conclusions 246
References 247

Glossary 251
Index 255

List of contributors

Syed Muhammad Zaigham Abbas Naqvi
College of Mechanical and Electrical Engineering, Henan Agricultural University, Zhengzhou, Henan, China; Henan International Joint Laboratory of Laser Technology in Agriculture Sciences, Zhengzhou, Henan, China; Department of Agronomy, PMAS-Arid Agriculture University Rawalpindi, Rawalpindi, Punjab, Pakistan

Farhat Abbas
College of Engineering Technology, University of Doha for Science and Technology, Doha, Qatar; Canadian Centre for Climate Change and Adaptation, University of Prince Edward Island, Charlottetown, PE, Canada

Hassan Afzaal
Faculty of Sustainable Design Engineering, University of Prince Edward Island, Charlottetown, PE, Canada

Aneela Afzal
Department of Sociology and Anthropology, PMAS-Arid Agriculture University Rawalpindi, Rawalpindi, Punjab, Pakistan; Department of Agricultural Extension, PMAS-Arid Agriculture University Rawalpindi, Rawalpindi, Punjab, Pakistan

Muhammad Naveed Anjum
Department of Land and Water Conservation Engineering, Faculty of Agricultural Engineering and Technology, PMAS-Arid Agriculture University Rawalpindi, Rawalpindi, Punjab, Pakistan; Data Driven Smart Decision Platform, PMAS-Arid Agriculture University Rawalpindi, Rawalpindi, Punjab, Pakistan

Muhammad Awais
Research Center of Fluid Machinery Engineering & Technology, Jiangsu University, Zhenjiangv, Jiangsu, China

Mark Bell
Strategic Initiatives and Statewide Programs, University of California: Agriculture and Natural Resources, Davis, CA, United States

Melanie Bos
Faculty of Sustainable Design Engineering, University of Prince Edward Island, Charlottetown, PE, Canada

Muhammad Jehanzeb Masud Cheema
Department of Land and Water Conservation Engineering, Faculty of Agricultural Engineering and Technology, PMAS-Arid Agriculture University Rawalpindi, Rawalpindi, Punjab, Pakistan; National Center of Industrial Biotechnology, PMAS-Arid Agriculture University Rawalpindi, Rawalpindi, Punjab, Pakistan

Richard A. Crabbe
School of Agricultural, Environmental and Veterinary Sciences, Charles Sturt University, Albury, NSW, Australia

Andre Daccache
Department of Biological and Agricultural Engineering, University of California, Davis, CA, United States

Karen Esau
Department of Engineering, Faculty of Agriculture, Dalhousie University, Truro, NS, Canada

Travis J. Esau
Department of Engineering, Faculty of Agriculture, Dalhousie University, Truro, NS, Canada

Aitazaz A. Farooque
Faculty of Sustainable Design Engineering, University of Prince Edward Island, Charlottetown, PE, Canada; School of Climate Change and Adaptation, Faculty of Science, University of Prince Edward Island, Charlottetown, PE, Canada; College of Engineering Technology, University of Doha for Science and Technology, Doha, Qatar; Canadian Centre for Climate Change and Adaptation, University of Prince Edward Island, Charlottetown, PE, Canada

Zainab Haroon
Data Driven Smart Decision Platform, PMAS-Arid Agriculture University Rawalpindi, Rawalpindi, Punjab, Pakistan

Patrick J. Hennessy
Department of Engineering, Faculty of Agriculture, Dalhousie University, Truro, NS, Canada

Fiaz Hussain
Department of Land and Water Conservation Engineering, Faculty of Agricultural Engineering and Technology, PMAS-Arid Agriculture University Rawalpindi, Rawalpindi, Punjab, Pakistan

Saddam Hussain
National Center of Industrial Biotechnology, PMAS-Arid Agriculture University Rawalpindi, Rawalpindi, Punjab, Pakistan; Department of Biological and Agricultural Engineering, University of California, Davis, CA, United States; Department of Irrigation and Drainage, University of Agriculture, Faisalabad, Punjab, Pakistan

Tahir Iqbal
Faculty of Agricultural Engineering and Technology, PMAS-Arid Agriculture University Rawalpindi, Rawalpindi, Punjab, Pakistan

Humna Khan
Department of Engineering, Faculty of Agriculture, Dalhousie University, Truro, NS, Canada

Mobushir R. Khan
School of Agricultural, Environmental and Veterinary Sciences, Charles Sturt University, Albury, NSW, Australia

Vijay Kumar
Department of Entomology, Punjab Agricultural University (PAU), Ludhiana, Punjab, India

Yubin Lan
National Center for International Collaboration Research on Precision Agriculture Aviation Pesticides Spraying Technology, South China Agricultural University, Guangzhou, Guangdong, China

Jana Levison
School of Engineering, University of Guelph, Guelph, ON, Canada

Craig B. MacEachern
Department of Engineering, Faculty of Agriculture, Dalhousie University, Truro, NS, Canada

Naeem A. Malik
Institute of Geo-Information & Earth Observation, PMAS-Arid Agriculture University Rawalpindi, Rawalpindi, Punjab, Pakistan

Lachlan O'Meara
School of Agricultural, Environmental and Veterinary Sciences, Charles Sturt University, Albury, NSW, Australia

Shoaib Rashid Saleem
School of Engineering, University of Guelph, Guelph, ON, Canada; Data Driven Smart Decision Platform, PMAS-Arid Agriculture University Rawalpindi, Rawalpindi, Punjab, Pakistan

Arnold W. Schumann
Citrus Research and Education Center, Institute of Food and Agricultural Sciences, University of Florida, Gainesville, FL, United States

Manjeet Singh
Department of Farm Machinery and Power Engineering, Punjab Agricultural University (PAU), Ludhiana, Punjab, India

Muhammad Naveed Tahir
Department of Agronomy, PMAS-Arid Agriculture University Rawalpindi, Rawalpindi, Punjab, Pakistan

Aseem Vermaa
Department of Farm Machinery and Power Engineering, Punjab Agricultural University (PAU), Ludhiana, Punjab, India

Yingkuan Wang
Chinese Academy of Agricultural Engineering Planning and Design, Beijing, China

Huang Wenjiang
Aerospace Information Research Institute, Chinese Academy of Sciences, Beijing, China

Ray-Shyan Wu
Department of Civil Engineering, National Central University, Chung-Li, Taiwan

Qamar U. Zaman
Department of Engineering, Faculty of Agriculture, Dalhousie University, Truro, NS, Canada

Yali Zhang
National Center for International Collaboration Research on Precision Agriculture Aviation Pesticides Spraying Technology, South China Agricultural University, Guangzhou, Guangdong, China

Foreword

The world now has over 8 billion people. Perhaps humanity's most challenging task is to provide all those people with sufficient food, feed, fiber, and fuel in changing climates while protecting the environment for future generations. The challenge is further heightened by the rightful continued development and increasing demands of less-developed populations and the need for economic, environmental, and sociopolitical sustainability.

One of the ways to help increase productivity, efficiency, and environmental protection is to account for the naturally occurring spatial, temporal, and weather variabilities. Using tools, techniques, and technologies to properly determine and respond to these variabilities is the crux of precision agriculture. This book therefore studies the evolution, insights, and trends of precision agriculture.

Many of us have been working in precision agriculture for 40 years or more. From small beginnings, there has been much work done and much achieved. It is therefore up to books such as this to summarize, organize, share, and transfer the knowledge gained. This book is a significant and substantial contributor to that sharing as it provides useful knowledge for students, researchers, extension professionals, and practitioners. It obviously cannot include all the knowledge from almost of half-century of work, but it has a good and well-chosen selection of what has been done and learned.

The editor of this book is Prof. Qamar Zaman who is in the midst of an excellent career in this area of work. He has a wide range of knowledge and experiences which are reflected in the diversity of the topics he has selected for coverage in this book. Besides the traditional topics covered in precision agriculture books, there is coverage of such topics as artificial intelligence, sustainability, irrigation, and the environment within the context of precision agriculture.

Another very positive aspect of this book is the diversity of the contributors to the book. The chapter coauthors hold professional positions in Australia, Canada, China, India, Pakistan, and the United States. These differences in geographic bases, climates, and agricultural practices lead to a variety of perspectives and inform the writing of the chapters.

The book starts with an introductory chapter by Prof. Zaman entitled "Precision Agriculture Technology: A Pathway Toward Sustainable Agriculture." The dozen following chapters expound upon that basis with coverage of specific topics of importance to precision agriculture. These chapters include discussions of the technologies, including such topics as UAV's, sensors, geographic information systems, remote sensing, big data, and artificial intelligence. They also include such topics as precision irrigation, variable rate technologies, sustainability and the environment, and technology adoption.

The chapter titles are properly descriptive and each chapter starts with a chapter outline and an abstract. It is therefore clear what will be covered in each chapter. The chapters discuss the literature in their respective areas and have good photos and illustrations. As usually is the case when reading about precision agriculture, the diagrams in some of the chapters are very useful.

Obviously, whole books could be written on any of the chapter topics and that level of detail is not possible in a single chapter. But the chapters in this book provide good introductions and good summaries of the various precision agriculture topics. The chapters are readable and do not depend upon the reader having advanced scientific or mathematical background knowledge. I therefore think it would be more suitable for students and practitioners rather than advanced researchers. At a minimum, it should be included in the libraries of colleges and universities which study agriculture.

John K. Schueller
Professor of Mechanical and Aerospace Engineering
Affiliate Professor of Agricultural and Biological Engineering
University of Florida

Preface

Burgeoning world population dictates higher quantity and quality of food. Even more importantly, population increase is coming from countries in South Asia and parts of Africa, that are relatively less developed. So affordability of this food is more important than ever before. Climate change inflicts pressure on agriculture inputs, thereby making it difficult for these very countries to stick to their conventional agricultural practices. We shall have to rethink and reimagine the agriculture. Centuries-old agriculture civilizations in valleys of Indus, Euphrates and Tigris, and Nile are finding it tough to reconfigure their growing practices. We embarked on developing this manuscript with a clear intention to employ technology to harness better quantity and quality of food and agriculture that too with minimal inputs utilized to the perfect precision. During the whole time, we were mindful to climate change so the technology employed in this book is geared to minimize agriculture impact on the environment. The planet implores mankind attention and sensitivity to the climate change so that blue waters sparkle is not lost, fish and aqua marine life's habitat is safe and thriving, glaciers maintain their natural life cycle, and rivers do not deviate from their centuries-old ebbs and flows. Let's save the mother nature, let's preserve the blue planet, and let the human race thrive with all its ingenuity. Let's embrace life.

List of abbreviations

$	Dollar
%	Percentage
3D	3-Dimensional
ACIAR	Australian Centre for International Agriculture Research
ACO	Ant Colony Optimization
AFS	Agriculture Farming System
Ai	Aphid index
AI	Artificial Intelligence
AKIS	Agriculture Knowledge and Information System
ANN	Artificial Neural Network
ASTER	Advanced Spaceborne Thermal Emission and Reflection Radiometer
ATV	All-Terrain Vehicle
AWS	Automatic Weather Station
AYMS	Automated Yield Monitoring System
BP	Backpropagation
C	Carbon
CCM	Color Co-occurrence Method
CEC	Cation exchange capacity
CGR	Crop Growth Rate
CH_4	Methane
cm	Centimeter
CNN	Convolutional Neural Network
CO_2	Carbon Dioxide
COD	Chemical Oxygen Demand
CPU	Central Processing Unit
CSM	Crop Selection Method
Ct	Cell State
DA	Discriminant Analysis
DBE	Doug Bragg Enterprises
DEM	Digital Elevation Model
DGPS	Differential Global Positioning System
DL	Deep Learning
DSS	Decision Support System
DSSAT	Decision Support System for Agro-technology Transfer
DSWI	Disease Water Stress Index
DT	Decision Tree
DTM	Digital Terrain Model
E	Experience
EC	Electrical Conductivity
ECa	Apparent Electrical Conductivity
ELM	Extreme Learning Machine
ELU	Exponential Linear Unit

EN	Elastic Net
ET	Evapotranspiration
ETLs	Economic Threshold Levels
ETo	Reference Evapotranspiration
EXIF	Exchangeable Image File Format
FAO	Food and Agriculture Organization
FC	Field Capacity
ft	Forget State
GCMs	Global Climate Models
GCS	Ground Control Station
GDP	Gross Domestic Product
GHG	Greenhouse Gas
GIS	Geographic Information System
GLONASS	Globalnaya Navigazionnaya Sputnikovaya Sistema
GMM	Gaussian Mixture Model
GNDVI	Green Normalized Difference Vegetation Index
GNSS	Global Navigation Satellite System
GPR	Ground Penetrating Radar
GPU	Graphical Processing Unit
GSD	Ground Sampling Distance
GST	Geospatial Technology
GWLs	Groundwater Levels
ha	Hectare
HBL	Habib Bank Limited
HCP	Horizontal Coplanar Geometry
HEIS	High Efficiency Irrigation System
HPS	High-Pressure Sodium
ht	Hidden State
ICT	Information and Communication Technology
ICWT	Inter Catchment Wastewater Transfer
ILSVRC	ImageNet Large Scale Visual Recognition Challenge
IoT	Internet of Things
IPCC	Intergovernmental Panel on Climate Change
IPM	Integrated Pest Management
IR	Infrared
IRR	Internal rate of return
ISO	International Organization for Standardization
IT	Information Technology
IWB	Water Band Index
IWO	Invasive Weed Optimization
kg	Kilogram
kNN	k-Nearest Neighbor
L	Liter
LAI	Leaf Area Index
LARS	Low Altitude Remote Sensing

LDM	Leaf Dry Matter
LED	Light Emitting Diode
LHI	Leaf Hopper Index
LIDAR	Light Detection and Ranging
LR	Linear Regression
LS	Least Square
LST	Land Surface Temperature
LSTM	Long Short-Term Memory
m	Meters
MARS	Mobile Agriculture Robot Swarm
MDGs	United Nations Millennium Development Goals
MGP	Mobile Ground Platform
ML	Machine Learning
MLP	Multilayer Perceptron
MODIS	Moderate Resolution Imaging Spectroradiometer
ms	Milliseconds
MSAVI	Modified Soil-Adjusted Vegetation Index
MSAVI2	Modified Soil-Adjusted Vegetation Index 2
MSE	Mean Squared Error
MSI	Moisture Stress Index
MTO	Met Station One
MZs	Management Zones
N	Nitrogen
N_2O	Nitrous Oxide
NDVI	Normalized Difference Vegetation Index
NDWI	Normalized Difference Water Index
NIR	Near-Infrared
NPQ	Nonphotochemical Quenching
NPV	Net present value
NUE	Nutrient Utilization Efficiency
OSAVI	Optimized Soil-Adjusted Vegetation Index
OSU	Oregon State University
P	Phosphorus
PA	Precision Agriculture
PAR	Photosynthetically Active Radiation
PAT	Precision Agriculture Technologies
PDI	Perpendicular Drought Index
PEI	Prince Edward Island
pH	Potential of Hydrogen
PPM	Precision in Pest Management
PPV	Plum Pox Virus
PRI	Photochemical Reflectance Index
PRZM	Pesticide Root Zone Model
RCP	Representative Concentration Pathway
ReLU	Rectified Linear Unit

RF	Radiative Forcing
RGB	Red, Green, and Blue
RMSE	Root Mean Square Error
rNIR	Near-Infrared Reflectance
RNN	Recurrent Neural Networks
RPM	Revolutions per Minute
rRed	Red Light Reflectance
RS	Remote Sensing
RTK	Real-Time Kinematic
RTK-GPS	Real-Time Kinematic—Global Positioning System
SAR	Synthetic Aperture Radar
SAVI	Soil-Adjusted Vegetation Index
SDGs	Sustainable Development Goals
SDI	Subsurface Drip Irrigation
SI	Satellite Imaging
SIRMOD	Surface Irrigation Simulation, Evaluation and Design
SSCM	Site-specific Crop Management
SSCMS	Site-Specific Crop Management System
SURDEV	Surface irrigation software for Design, operation and EValuation of basin
SVM	Support Vector Machine
SVR	Support Vector Regression
T	Task
TCT	Tasseled Cap Transformation
TDM	Total Dry Matter
TDR	Time-Domain Reflectometry
TH	Temperature Harvested
Th	Time at harvest
TIR	Thermal Infrared
TM	Thematic Mapper
TVI	Transformed Vegetation Index
UAS	Unmanned Aerial System
UAV	Unmanned Aerial Vehicles
URT	Uniform-rate technologies
US	United States
USA	United States of America
VIs	Vegetation indices
VIS	Visible
VIS-NIR	Visible and Near-infrared Spectral Sensors
VLR	Vertical Looking Radar
VR	Variable Rate
VRA	Variable-Rate Application
VRAM	Video Random Access Memory
VRNS	Variable-rate Nutrient Application
VRSS	Variable-rate Spraying Systems
VRT	Variable-Rate Technology

WI	Water Index
WP	Water Productivity
Wp	Wilting Point
WSN	Wireless Sensor Network
WSVI	Water Stress Vegetation Index
WUE	Water Use Efficiency
WUR	Wageningen University & Research
XOR	Exclusive OR
YOLO	You Only Look Once
Ψ	Soil Water Tensions or Matric Potential
θAW	Adequate Water Available
μEye	Micro-eye
ZTBL	Zarai Tarqiati Bank Limited

CHAPTER 1

Precision agriculture technology: a pathway toward sustainable agriculture

Qamar U. Zaman

Department of Engineering, Faculty of Agriculture, Dalhousie University, Truro, NS, Canada

1.1 Introduction

Precision agriculture is a part of the third wave of the agricultural revolution to achieve sustainability in agricultural productivity. The first wave comprised farm mechanization from 1900 to 1930 when every farmer could produce enough food for 26 people. The year 1960 witnessed a second wave termed as the green revolution supported by genetic modification enabling every farmer to feed 155 people. With a growing world population (to be 9.6 billion by 2050), every farmer must produce food for 260 people to ensure food security and promote sustainability. This would be possible using precision agriculture technologies. In the past, precision agriculture was addressing soil and crop variability issues. Presently, it is the use of advanced technologies for applying site-specific agricultural inputs to optimize crop yields. Its future endeavors envisage benefiting from artificial intelligence technology for enhancing farm profitability while protecting environmental conditions.

1.1.1 Background

Agriculture has been practiced over ages to sustain the living creature on Earth including humans by producing fruits, vegetables, and other byproducts for their consumption. It is an essential resource for mankind's survival due to its capability of providing food and fiber. It involves traditional and modern farming practices including soil management practices to grow various crops and animal husbandry for food, yarn, and poultry products for the growing population. By 2050, about 60% of the total world's population will depend directly or indirectly on agriculture for earning their livelihood [1], which requires much attention to overcome shortages of the agricultural lands in view of mushroom growth of the housing societies.

Other than farming practices for growing crops, agricultural activities also involve raising livestock, poultry, and fish for meat, milk, and eggs. On the average, red meat, poultry, seafood, and processed meat (e.g., sausages, frankfurts, saveloy, bacon, ham, fermented, comminuted meats, processed delicatessen meat) are consumed at the rate of 22.7, 21.1, 9.49, and 7.99 kg/year/capita [2]. For building immunity against chronic diseases, the daily consumption of fruits and vegetables is recommended to be 0.4 kg/person [3]. For 6.453 billion people out of the 7.753 billion population of the world minus 1.3 billion children under age of 14, the requirements for fruits and vegetables become 2.58×10^6 Mg/year. For these facts and figures, and mankind's survival on Earth, agriculture will remain the backbone for the life sustenance forever.

1.1.2 Footprints of agricultural practices and products

Similar to the impressions or images that humans or animals leave while they pass through an unpaved field, past agricultural practices have left their footprints on the ecosystem. These may include water and carbon footprints. The water footprint includes the amount of water used to produce a food product including meat, fruit, vegetables, grain, and/or their byproducts. The carbon footprints are quantities of the greenhouse gases emitted during anthropogenic activities or life cycle of the agricultural products. The water and carbon footprints can be quantified for individuals, groups, or organizations per unit of the product (weight or volume unit of food like kilograms of grains or liters of milk) production and/or per unit of time (per day, per growing season, or year). These facts stipulate sustainability of the agricultural production system for ensuring food security for the nation.

1.1.3 Need for agricultural sustainability

To meet the food security challenges of the growing world population, agriculture practices need to promote sustainably. This necessarily means contributing to the achievements of the United Nations Millennium Development Goals (MDGs) and Sustainable Development Goals (SDGs). SDG-2 targets alleviating hunger, attaining food security and improving food nutrition by 2030 through practicing and promoting sustainable agricultural practices. Combating extreme poverty and hunger challenges is among MDGs in addition to reducing environmental degradation during this millennium. The environmental sustainability goal (MDG-7) includes reversing the loss of environmental conditions. Depending on the available resources and commitment to the cause, and seriousness in the binding of international accords, these global targets can be met region by region, country by country, and on an individual basis. Humans and animals including tiny creatures can only survive if humans strive for meeting the needs of the current generations (while conserving the natural resources and protecting the global ecosystems) without challenging the requirements of the future generations for their ability to meet their own needs. Sustainable agricultural practices can improve the quality and quantity of food, drive economic revolutions, and enable the world's deprived people to avoid poverty in the context of the looming crisis of climate change impact on food security issues.

1.2 Climate change, food security, and precision agriculture nexus

The nexus of climate change, food security, and precision agriculture is important for understanding in order to combat threats of climate change, food insecurity, and inefficient conventional agricultural practices on food security.

1.2.1 Threats of climate change

The conventional agriculture contributes about 25% to the total greenhouse gas emissions worldwide. Whereas environmental pollution is caused by a variety of factors including emission of the greenhouse gases that not only give rise to the atmospheric temperatures but also enhance cases of diseases. Climate change is not only considered as a simple environmental problem but it also poses the greatest threat to the global food security issues. Its devastating effects include continuous rise in global temperatures, the resultant accelerated melting of glaciers, consequential rise in sea level, loss of essential biodiversity, occurrence of the extreme weather events, floods, and droughts, and global outbreaks of the fatal diseases [4].

1.2.2 Food security challenges

According to the World Food Summit (1996), "Food insecurity exists when all people, at all times, do not have physical and economic access to sufficient, safe and nutritious food that meets their dietary needs and food preferences for an active and healthy life." In turn, food insecurity leads to poor physical and cognitive development and ultimately to low productivity and poverty. Therefore, for a country to be food secure, its people must ensure to have adequate, safe, and pure food availability for leading an active and healthy life. The food insecurity experience scale of FAO is shown in Fig. 1.1, according to which 26% of the people worldwide face food security challenges and 11% of people face severe food insecurity issues [5].

1.2.3 Solutions embedded in precision agriculture adoption

Precision agriculture adoption embeds in itself the solutions for food security, climate change mitigation, and per hectare yield increase. Mueller et al. [6] claimed to meet the global yield gaps by 79% (equivalent to an increase in global food production by 29%) through practicing precision agriculture technologies. Such higher crop yields will ensure food security and farm profits. The environmental benefits of precisions agriculture include achieving higher crop yields with less water and carbon footprints, no expansion of farmlands, reduced deforestation, and natural resource conservation. The minimum possible use of agrochemicals and soil conservation techniques will cause a reduction in greenhouse gas emissions, soil contamination, water pollution, and the spread of certain epidemics.

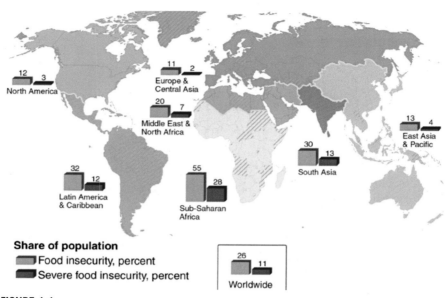

FIGURE 1.1

FAO food insecurity experience scale and USDA findings [5].

1.3 Agricultural resources management

Soil, water, nutrients (organic and inorganic), light, and atmospheric carbon dioxide (CO_2) are included among major agricultural resources. For field agricultural activities, the soil is sourced from land that also facilitates farm structures including buildings to store equipment, process compost, and conduct postharvest handling and processing of the agricultural produce. The farmlands also host activities/offices to market products of processed crops, livestock, and livestock products. Agricultural resources management aims at addressing challenges caused by the field spatial variability effects being faced by the farmers.

Factors such as soil moisture, electric conductivity, topography, and texture are the key variables to differentiate various regions and fields. Furthermore, crop variability tends to include crop height, density, nutrients, and water stress [7]. The crop height and density are the physical features that can be observed whereas the nutrient and water stress tend to be found by using various sensors/analyses [8]. Variability in anomalous factors can be defined as any randomly occurring impacts on the agricultural fields [9]. This can include a few examples such as weed infestations, insect infestations, wind damages, crop diseases, and frost [7]. Another factor of variability that leads to a large impact on crop production is management variability which is due to human decision error in the farming practices that can be a result of crop seeding rate, crop rotation, inadequate application of fertilizer and pesticide, and irrigation patterns [10]. The focus of the precision agriculture technologies is optimum use of the agricultural resources, that is, land, water, fertilizer, and management practices to promote sustainability of these resources, increase farm profitability through increasing crop yields as well as protect the environment by making lesser use of the agrochemicals.

1.3.1 Soil management

Soil management on the farmlands is mainly practiced to protect soil potential while practicing soil conservation techniques during seedbed preparation of new crops to reduce soil erosion and increase soil structure stabilization. Zero or no-till, minimum tillage, conservation tillage, and shallow tillage are practiced as efficient soil management practices in comparison with deep tillage except when the deep tillage is required to break the soil's subsurface hardpan. Moreover, crop residues can be left on the soil surface for their incorporation into the soil by cultivation to enhance soil fertility through increasing organic matter contents, which help improve soil structure as well as retain soil moisture for better crop growth.

1.3.2 Nutrient management

Soils are supposed to provide the necessary nutrients and minerals for adequate plant growth. Therefore, soil nutrient management is directly related to soil management to ensure fertile and healthy soils for optimum crop yields. Soil's major and minor nutrients are derived from organic and inorganic fertilizers. Management of organic fertilizer does not only help manage livestock's byproducts to avoid methane (CH_4) emissions but also increases soil organic matter, which is vital for the soils to remain productive as it holds soil water and retains nutrients making them available for plant uptake. In contrast to inorganic fertilizers, organic fertilizers carry the least risk for environmental contamination (i.e., soil and water (ground and surface) pollution). The pollution, however, originating from the leaching of the nutrients below the crop root zone can be avoided by efficient management of the irrigation water applications. Light, frequent, and precise irrigations using high-efficiency irrigation systems can minimize the leaching of the nutrients out of the root zone.

1.3.3 Water management

The adequate management of water resources including soil water involves maintaining the soil water contents at an optimal state in the root zone to fulfill the actual crop water requirements. Irrigation scheduling is practiced to replenish the soil water requirements for the crops. The schedule for applying supplemental irrigation water to the plants to maintain soil moisture between the limits of their soils' wilting point (WP) and field capacity (FC) is called irrigation scheduling. In other words, irrigation scheduling guides about the time and amount of irrigation water to maintain adequate water available (θ_{AW}) for plants. It is the available soil water contents between wilting point (θ_{WP}) and field capacity (θ_{FC}). The soil FC condition quantifies the soil water availability for plant use after all gravitational drainage has been completed. The soil wilting point is the amount of soil water that is least available for plant uptake. The soil water at the permanent wilting point (a stage next to θ_{PW}) is not completely dry though, but the amount of water in the soil is so little that it is not available for plant usage.

Irrigation should be applied to the plants to maintain θ_{AW} (i.e., $\theta_{AW} = \theta_{FC} - \theta_{PW}$) by not exceeding θ_{FC} and not depleting below θ_{PW} upper and lower limits, respectively. The soil water content values for θ_{FC} and θ_{PW} of different types of soils can be extracted from values of soil water tensions or matric potential (Ψ) for different soils which are -0.1 to -0.33 bars for θ_{FC} and -15 bars for θ_{PW} [11]. Soil water characteristics curves ((i.e., relationships between Ψ in kPa (1 kPa = 1 cbar) with θ (%)) can be constructed for various types of soils and growing media. For peat soils, the soil water characteristic curves are flatter than the curves for sand, clay, and clay loam soils showing better soil water withholding capacity.

The scientists are of a mixed opinion about the relationships between Ψ and θ. The sandy soils retain 15%−25%, loamy soils retain 35%−45%, and clayey soil retains 45%−55% θ at FC. Similarly, about 33% θ is retained by the fine-textured soils, and 10%−15% θ is retained by coarse-textured soils. For various soils, the values of θ_{PW} and θ_{FC} can be calculated at a Ψ of -1500 cbars and -30 cbars, respectively. The case of peat soils, however, is different. Norrie et al. [12] designed an irrigation system that was automated for greenhouse use to grow tomatoes in peat-based growth media for measuring matric potential (water tension) in the growth media. They demonstrated that crop irrigation requirements can be accurately determined using electronic tensiometers to determine irrigation starting times. They used -20 cbar matric potential to start and -5 cbar matric potential to stop irrigation. This converts to ~ 30% and ~ 65% soil water contents in the peat bags to be considered as θ_{PW} and θ_{FC}, respectively. Cormier et al. [13] also used -5 cbar as the threshold for peat moss cultivated-tomatoes irrigation in a greenhouse study. The high irrigation threshold, i.e., -5 cbar for peat moss is because of their surface area and porosity that cause low availability of soil water for plants to use along with the very limited volume of the plant roots structure developed therein [14]. The work of Jobin et al. [15] and Caron et al. [16] also confirms this characteristic of the peat moss.

1.3.4 Supplemental light and CO_2

Light and CO_2 are naturally available resources for field agricultural requirements. They, however, are artificially managed for greenhouse agricultural practices. Their supplemental and artificial management in a greenhouse can help achieve a higher quality yield on a sustainable basis [17]. The red and blue greenhouse lights with 400 to 700 nm wavelength range positively influences the photosynthesis processes [18].

The light wavelength ranging from 400 to 700 nm is visible to the human eyes. It can be divided into the ranges of 380−499 nm having a blue spectrum, 500−599 nm with a green spectrum, and 600 to 700 nm having a red spectrum. Different plant species use various light spectrums for optimum growth and yield. For example, tomato (*Solanum lycopersicum* L.) plants grown under LED (light-emitting diode) have lesser photosynthetic capacity compared to those grown under HPS (high-pressure sodium) lamps as the HPS lamps stimulate more generation and early production [19,20].

McCree [21] evaluated a light environment system to grow crops under red and blue wavelengths and found that the upper limits of the tested lights produced better photosynthesis results than those under lower portions of these lights. They considered 600–700 nm of red and 400–500 nm of blue light wavelengths as the most photosynthetically efficient spectral range. Deram et al. [14] used three red and blue (661 and 449 nm, respectively) ratio levels (i.e., 5:1, 10:1, and 19:1) provided by LED and HSP to grow tomatoes in a greenhouse. They found a red to the blue ratio of 19:1 to produce the highest biomass. The red to blue ratio of 5:1 produced better fruit. The LED: HPS (50%:50%), however, produced better marketable fruits over 90 g weight followed by 5:1 high and 19:1 treatments. Nguyen et al. [17] cultivated tomatoes with 436 nm blue and 526 nm green high wavelengths in a greenhouse and found that these blue and green light wavelengths increased the yield of tomatoes by about 73% and 71%, respectively, than the arrangements of their experiment under natural light considered as a control treatment.

Hemming et al. [22] recommended a supplemental lighting system for a 96 m^2 floor area greenhouse consisting of six HPS lamps. These lamps were of 100 μmol/m^2/s capacity received from 1000 W light bulbs. Their experimental treatments also involved eight LED lamps that had multispectrum capacity. The capacity was controllable from 0 to 109 μmol/m^2/s. The LED treatments had mixed spectrums of blue, red, and white lights having capacities of 11, 49, and 37 μmol/m^2/s, respectively. According to OSU (Oregon State University) normal tomato foliage growth requires a minimum of 650 foot candles of light. About 600 foot candles of light can be yielded from a 1000 W metal halide for a 112 ft^2 area of the greenhouse. The use of different light spectrum ranges and their different mutual ratio have shown significant effects on plant growth within greenhouses, which can be improved further using precision agriculture technologies with automated sensing and application system.

1.4 Precision agriculture tools, techniques, and technologies

A driving factor of many negative impacts of conventional agriculture would be the use of agrochemicals, which can be minimized using precision agriculture technologies. Recently, farmers are under pressure to meet the world's food requirements which are increasing with the consistent increase in the world's population. This leads to the use of agrochemicals to optimize their crop yields, which induce damaging effects on the surrounding environment. Some of these damaging effects include subsurface groundwater contamination, nutrient runoff, greenhouse gas emissions, and volatilization.

Subsurface leaching of water, resulting from the basin and/or over-irrigation of agricultural fields is a disadvantage of the conventional agricultural practice, and thus nutrients leach down from the agricultural fields. Such soil and groundwater contaminations occur with the regular accumulation of soluble salts, pathogens, pesticides, and nutrients, which can easily be concentrated with irrigation or rainwater and transported down to the groundwater. Nitrate leaching is an example of one of many forms of subsurface leaching. Nitrate leaching is a common result of the abundant use of nitrogenous fertilizers, although nitrogen is an essential element or nutrient for plant growth. The excessive use of nitrogenous fertilizers also stimulates its loss to the atmosphere through volatilization. The main concern of this subsurface water contamination is the presence of the high levels of pathogens and heavy metals. To mitigate these negative effects of the conventional agricultural practices, precision agriculture can be adopted to ensure food security along with environmental protection for the overall well-being of our planet.

Keeping in view the above environmental repercussions as a result of offsite transport of agrochemicals from the agricultural fields to the environment, i.e., water bodies, use of the site-specific or precision agriculture technology was made to minimize the offsite transport/leaching of the agrochemicals. As precision agriculture technology makes use of the spatial variability effects of the soil potential on the crop yields. Some parts of the field are more productive compared with others by virtue of variability in the soil fertility. Moreover, soil properties vary from point to point in the field and thus affect crop growth accordingly. Precision agriculture advocates a variable rate of application of agrochemicals, and fertilizers to produce uniform crop growth across the field by applying more fertilizer to nutrient-poor soils and less fertilizer to the nutrient-rich soils. The overall objective is to minimize the use of agrochemicals while getting better yields and protecting the environment. For this purpose, there is a need of mapping the soil properties such as texture, nutrients, and topography with the help of DGPS (Differential Global Positioning System), GIS (Geographic Information System), computer simulation models along with sensors for delineating zones to apply variable rate technology. The advent of advanced technologies in these areas of DGPS, GIS, sensors automation coupled with software has enabled the variable rate applicators to apply agrochemicals on the go as per the requirements of the soil and crops. The use of these precision agriculture technologies and techniques has resulted in lesser use of agrochemicals, thus protecting the environment and producing better crop yields and farm profitability.

1.4.1 Precision agriculture

Precision agriculture practice makes use of the spatial variability effects of the soils to apply site-specific/variable rate agricultural inputs for enhancing farm profitability while protecting the environment. To apply the site-specific variable rate inputs, mapping of the soil properties as well as crop yields is prepared using GPS, GIS, and computer modeling technologies to delineate the homogenous zones based on the crop yields variability for applying variable rate agricultural inputs accordingly to minimize the use/cost of the inputs. Precision agriculture promotes lesser use of chemicals/inputs, thus reducing the cost of production, enhancing crop yields, as well as protecting the environment. It uses farm resources in a conservative but precise way to avoid resource wastage and ensure optimum crop yields. It enables the farmers to maximize their farm output by use of the minimum possible inputs. It recycles and reuses the farm products thereby reducing waste under the principles of a circular bioeconomy that is powered by nature. For example, livestock waste such as cow dung, biomass waste such as wheat straw or rice husk, and energy produced from the processes of transforming livestock and biomass waste to compost and biochar, respectively, for farm structure heating. Practicing precision agriculture requires a range and variety of tools, techniques, and technologies for performing environment-friendly farm operations efficiently as briefed below.

1.4.2 Precision agriculture tools

Machines such as tractors, plows/tillers, seeders, sprayers, pruners, harvesters, fruit and vegetable pickers, threshers, cranes, trailers/trollies, loaders, dumpers, and haulers are some examples of precision agriculture tools when equipped with the GPS units and monitoring mechanism. Equipment such as global positioning systems (GPS), soil moisture sensors, electrical conductivity sensors, greenhouse gas monitoring devices, soil temperature sensors, plant-canopy photosynthetically active radiation (PAR) sensors, plant leaf sensors to measure chlorophyll content, leaf moisture, leaf indices including NDVI (normalized difference vegetation index) and smart screen phones equipped with sensors and cameras to detect the plant temperature, plant-water, disease, and pest stresses, and plant species are also listed as precision agriculture tools. They further include drones and cameras installed on the drones for thermal mapping of an agricultural field and yield monitoring. The availability of modern tools does not overturn the importance and use of basic and conventional farm tools used for a variety of tasks that cannot be performed mechanically or economically. These tools include sickles, hoes, machetes, and other gardening tools used for trenching, harvesting, weeding, hoeing, and other field and/or gardening tasks.

For irrigating crops, trees, or individual ornamental plants at home or in greenhouses, the list of precision agriculture tools is not complete excluding pumps, motors, pressure irrigation system (drip or sprinkler) tools including water storing tanks, main and lateral water lines, heads of sprinklers and drips, and all tools used in a weather station including rain gauges, sensors for humidity and solar radiations, and wind vanes to measure wind direction equipped with anemometers used to measure wind speed.

Maximum output from livestock farming is possible through adopting proper animal husbandry practices. The best health and comfort of animals result in their satisfaction and thus optimum output in terms of meat, milk, and eggs. Livestock management practices require the use of certain tools. They include hay racks for feeding, troughs for goat feeding, relaxing, and eating, self-feeders equipped with feed hoppers to automatically supply food to livestock, bowls to stock feeding racks and troughs, shovels/spades to mix feeds, water troughs, and shades to keep farm animals cool during summer, buckets and hoses for animal drinking, dipping vats for cattle swimming, incubators for the hatching of female birds, and brooders to provide heat to the young poultry birds. As livestock farming is an important pillar for ensuring food security; however, the use of precision agriculture technologies has potential to further enhance the economic returns from animal husbandry/livestock farming.

1.4.2.1 Geographical information system
One of the useful precision agricultural tools is GIS (Geographic Information System). It helps farmers in making informed decisions based on geographic/position data. The field data are analyzed to create visual representations of the spatial properties of the fields. It comprises data, software, and hardware whereby the data of soil or crop properties are collected along with the geographic components/coordinates (northing and easting) of the data points using GPS units. Special software comprises various applications, which are used to analyze the collected spatial data using computer software/hardware systems. Data collection tools such as hand-held, ground, or motor-mounted GPS units, proximal sensors, drones, and satellites are the external units of the GIS computing system that comprises desktops, laptops, or pads. ArcMap, ArcInfo, ArcView, ArcCatalog, and ArcEditor are some of the core applications of ArcGIS software. GIS is used to map temporal and spatial data for detecting soil or crop variability that is used as the basis for variable-rate agricultural inputs under precision agricultural technologies.

1.4.3 Precision agriculture techniques
Precision agriculture is itself a technique of modern intelligent farming system in contrast to conventional farming which uses old, basic, and inefficient tools and technologies. The old-style conventional farmers use basin irrigation without irrigation scheduling ending in water losses resulting in low water use efficiency. Intelligent farmers use pressurized irrigation systems including sprinkler or drip irrigation systems to enhance crop water productivity. The conventional farming practices include uniform application of agrochemicals without considering soil and/or crop variability potentials. Precision agriculture promotes variable rate application of agrochemicals and seed spacing based on recommendations from science-based experimentations or through real-time data collection systems.

Crop yield losses are significantly higher under conventional farming due to the use of manual harvesting practices with the conventional tools, and one-time harvesting of fruits and vegetables when crops are either not fully ripened or are overripened. The use of precision agriculture technologies does not only help reduce crop yield losses, but also results in maximizing farm income due to the benefits of automation and the least use of human labor.

1.4.3.1 Artificial intelligence

Artificial intelligence has helped improve precision agriculture practices in numerous ways. For example, communication between and among a network of agricultural sensors is made possible with algorithms of machine learning and the internet of things (IoT). Efficient networking of IoT sensors can improve agricultural equipment operations and farm efficiencies, enhance crop yields, save the use of manhours, reduce food production costs, and thus optimize farm profits.

The use of artificial intelligence has eased the efforts of farm operators. The calculations made conventionally with the use of calculators, papers, and pens were conveniently transformed to be performed at a large scale by using Excel spreadsheets previously. Artificial intelligence is now helping farm operators beyond the use of Excel spreadsheets, that is, in gaining insight into meteorological variables, biodiversity at the farm, accurate application of agrochemicals, management of intercropping, soil conservation, supplemental irrigation, and other internal as well as external factors affecting crop yields and product quality. In summary, artificial intelligence can improve efficiencies of precision agriculture practices by:

- Video surveillance and detection of objects/subjects interfering with the farm operations.
- Identification of intruding animals that cause potential damage to the crops and farm structures.
- Crop yield prediction for timely arrangements of harvesting and estimation of harvest losses to mitigate the expected damages.
- Well-equipped agricultural machinery has transformed farming systems to a new level of smartness.
- Crop productivity, interactive tracking, efficient harvesting, processing, and real-time marketing have become a reality.
- Detection and tracking of outbreaks of weeds and crop diseases have become possible with machine vision.
- Artificial intelligence has helped promote modern and more efficient agricultural methods such as a greenhouse or controlled farming to grow daily-use fruits and vegetables.
- Agricultural operations have become unmanned with the use of machine vision, deep learning, and artificial intelligence.
- Accompanied by IoT, artificial intelligence supports image capturing with proximal sensing techniques such as drones and remote sensing with satellites, image processing, and image analysis to help farm managers monitor their agricultural fields, operate agricultural machinery, and operate control systems remotely.
- In livestock farming, unmanned aerial vehicles can assist farm managers in tracking the wondered farm animals, sick and/or wounded sick animals, and herd count per grazing using sensor and GPS technologies.

In a nutshell, the role of artificial intelligence has become unbelievably intelligent and unavoidably crucial in the field of precision agriculture technologies.

1.4.4 Precision agriculture technologies
1.4.4.1 Variable-rate application technologies

Precision agricultural technologies are based on soil or crop variability data. A grid pattern is established at the field to collect geocoordinated soil or crop variability data using precision agricultural tools. For example, the DualEM-II sensor can be used to measure ground apparent electrical conductivity (ECa), TDR-300 can measure soil moisture (θ) content and a soil penetrometer is used to diagnose the soil compaction data. The data can then be analyzed and mapped spatially to build a relationship between ECa, θ, and cone index (soil resistance to penetration) data. Now, the farmers do not need to prepare a cone index for breaking the subsurface hardpan but run DUALEM to map ECa only and use the ECa-θ-cone index relationship to determine spatially distributed subsurface soil hardpans. The map and the relationship can be received automatically by the control system of the mouldboard plow for deep tilling to break the site-specific soil compacted layers.

Artificial intelligence can be used for developing and implementing site-specific optimal algorithms for agrochemicals through variable-rate applicators for enhancing farm profits, environmental protection, and climate change mitigation. The rationale behind this approach is that a uniform-rate application of agrochemicals is used in conventional agriculture, without considering the variations (e.g., spatial and temporal) in the soil, crop, weather, and topographic attributes that exist within the fields. Such uniform application of agrochemicals results in increasing farm production costs and causes adverse impacts on the environment in the shape of high GHG (greenhouse gases) emissions, leaching of nitrates to the groundwater, and nutrient runoff to the water bodies posing risks to the environment and the aquatic life. One would end up with increased use of resources, adverse environmental effects, posing danger to the aquatic life, causing higher GHG emissions, aggravating the existing climate change impacts and global warming by opting for conventional agriculture for food security efforts when compared with the precision agriculture practices. Site-specific applications of agrochemicals with variable-rate sprayers can save farmers' resources, and protect environment, and planet Earth from further warming.

The Researchers have developed an automated variable rate spraying system to apply agrochemicals as and when needed to reduce crop input costs, control environmental degradation, minimize yield losses, and risks of global warming [23,24]. Farooque et al. [25] used a deep learning-based variable-rate sprayer for on-the-spot application of agrochemicals and concluded that the variable-rate sprayers are the suitable smart and precise applicators of agrochemicals at agricultural fields resulting in optimized use of agrochemicals, enhance plant protection, environmental safety, and high farm profits.

1.4.4.2 Seeding technology
Conventionally crop seeding has been done mostly manually. The next step was to invent seeding devices to be used manually. A commonly available handheld

instrument has small openings to pass seeds one by one. Improvements in handheld instruments brought mechanically working seeding devices to perform at higher precision. These devices are required to be calibrated for each type of seed, based on the respective standard diameter. Innovations in the development of precision agricultural technologies have introduced automated seeding machines to plant seeds and seedlings of various crops called rice seeding machines or rice seedling transplanters. The rice seeding machines are becoming popular and a big attraction for rice-growing farmers worldwide.

1.4.4.3 Satellite imagery for field mapping with remote sensing
The healthy plants exhibit low red-light reflectance (rRed) as their chlorophyll absorbs red light. Instead, the healthy plants reflect higher near-infrared (rNIR) light. Likewise, the weak plants or plants under stress have increased rRed and decreased rNIR. These variables are used to measure NDVI using remotely sensed rRed and rNIR measurements as NDVI = (rNIR − rRed) × 1/(rNIR + rRed). NDVI measurements indicate plant stress to crop inputs including nutrients and will be used to determine nitrogen deficiency in potato plants. Potato growing fields (with and/or without plants) will be sensed through remote sensing by capturing the color and thermal images of the fields using unmanned aerial vehicles (UAVs). The images are processed and used by GIS field mapping for showing the effects of soil health on crop quality and productivity. Such field mapping is useful in the estimation and control of soil erosion caused by climate change impact to improve crop productivity and profitability using precision soil conservation practices.

Crop spectroscopy/Sensing is used to determine leaves' photosynthetic activities. The use of UAVs, such as drones, has made plant and soil sensing possible. The UAVs use color and thermal cameras to capture images of agricultural fields. These Remote Sensing images can be related to the crop yield maps for determining the impact of soil health on the growth of plants or to determine the soil health from conditions of plant growth and crop yields.

1.4.4.4 Soil and crop proximal sensing
Proximal sensing is used to determine the soil and crop properties such as soil fertility and crop yield variability. The data collected with the proximal sensing approach can then be analyzed using machine learning algorithms to determine relationships between independent variables (e.g., soil and crop growth properties) and the dependent variables (e.g., crop yield). Abbas et al. [26] exploited the potential of precision agriculture technologies including proximal sensing of the soil and crop combined with data processing using machine learning (ML) algorithms to extract useful information about independent variables that regulate crop yield productivity. Various ML algorithms such as elastic net (EN), linear regression (LR), support vector regression (SVR), and k-nearest neighbor (k-NN) can be used to predict the potato tuber yield from the data retrieved using proximal sensing of the soil and crop properties. The independent data included for ECa, θ, soil slope, NDVI, and selected soil chemical properties. They reported that the SVR model had better performance

than that of the other models used in their study for one set of data. Such an approach can be used for designing and implementing the site-specific management zones approach for better crop productivity.

1.5 This book

This book comprises 13 chapters. Efforts have been made to include evolution, insights, and new trends in precision agriculture technologies. After this introductory chapter, Chapter 2 presents solutions and examples to manage the soil spatial variability as there always exists variabilities (e.g., spatial and temporal) in the physiochemical properties of the soils and crops. Such variabilities provide the basis for variable-rate application of agrochemicals. Chapter 3 is about Integrated Pest Management (IPM) and considers the variability of pests and diseases in agricultural fields to use the geospatial technologies for IPM.

Unmanned aerial vehicles are unpiloted flying robots and are commonly termed drones, micro-, and/or nano-air/aerial vehicles. Chapter 4 is about the use of UAVs in agriculture. Chapter 5 is about the applications of geospatial technologies for precision agriculture, which is believed to keep the environment safe, minimize climate change effects, and improve the food security situation. All of these aspects of sustainable agriculture are based on precision irrigation applications. Chapter 6 presents the challenges and opportunities for precision irrigation applications. This chapter provides information about different traditional irrigation methods and emerging precision irrigation systems. Precision irrigation is practiced as a variable rate application technology as it addresses the spatial and temporal variability in the soil moisture contents. The variable rate technologies have been developed in recent years for various cropping systems to replace the traditional uniform application of agrochemicals as well. Chapter 7 is all about variable rate technologies development, adaptation, and their opportunities in agriculture. The use of variable rate technologies ends up ensuring farm profitability. Another technology responsible for improving farm income is yield monitoring and precision harvesting. Chapter 8 explains how various factors affect the harvesting efficiency of the harvesters resulting in varying harvesting yields and losses. This chapter presents external factors (i.e., plant height, plant density fruit diameter, and field topography. Meteorology-related factors include harvest temperature, relative humidity, plant canopy wetness, and soil moisture) and the harvester-related factors (i.e., type and placement of sensors used on-the-go, ground speed, head revolutions, head diameter) and presents information about precision harvesting technologies to optimize the harvesting crop yield and minimize harvesting losses.

Like many other fields, precision agriculture has started benefiting from artificial intelligence and deep learning applications. Chapter 9 covers machine learning applications in agriculture to help increase efficiencies at the farm level and Chapter 10 covers the use of artificial neural network modeling for precision agricultural water management practices. A gentle introduction of the commonly used machine learning algorithms and libraries is discussed to familiarize the readers with recent advances in this specific domain. Important steps from training to implementation of machine learning algorithms in hydrological studies are also discussed.

Unlike other areas of business, agriculture has not been using advanced techniques of marketing and profit making. To cover this aspect of agriculture, Chapter 11 focuses on the economic and resultant social benefits of precision agriculture practices and reshaping the agricultural value chains. Chapter 12 presents the fact that conventional agriculture has remained one of the main contributors to environmental pollution during the past few decades. For example, the overapplication of agrochemicals has polluted various components of the environment such as surface water, groundwater, and soil. This chapter discusses the positive impacts of the best agricultural management practices on the environment for various cropping systems. Chapter 13 offers an outline for the development of the precision agriculture technologies for environmental safety, crop yield, farm profits, food security, and sustainability and their current status based on the published literature over the past two decades.

1.6 Conclusions

Given the scope of this book and the importance as well as acceptance/adaptation of precision agriculture technologies, it is believed that the book will attract a diverse set of audiences from practitioners to policymakers and researchers to students. The thrust of the book is not only on technology development but on working out the economics of the technology adoption. The focus on the economic, commercial, and environmental sides of precision agriculture—the topics that have been relatively less explored in the existing literature, will attract more audiences. The book provides a complete guide to precision agriculture practitioners from start to end including its economic, commercial, and environmental implications. Its contents are simplified to engage less technical and nontechnical readers to help promote precision agriculture practices, especially in those sectors which are not receptive to the adoption of the new technologies.

Several chapters of this book include, in addition to GIS and its applications, spatial variability, and precision agriculture as a management tool using remote sensing, expert systems, artificial intelligence, and the latest technologies and approaches. The discussions with the potential contributors have revealed that precision agriculture techniques, tools, and technologies used in the past have been improved during the present research era, and future trending, as well as innovatory use of artificial intelligence to promote sustainable agriculture, are covered in this book. Some of the key benefits of this book to the readers include the following points:

(i) Agricultural practitioners are mostly undecided about whether to invest in the technology or not. This book steers them through a financial sense of investing in this precision agriculture technology.

(ii) Once decided to adopt Precision Agriculture, the book provides a comprehensive guideline of contemporary and future technologies in the field so that readers can make an informed decision regarding the obsolescence of a particular precision Agriculture technology.
(iii) Comprehensive cost-benefit analysis has been included in the book for governments and other policymakers to incentivize and promote precision agriculture technology.
(iv) Detailed reviews of precision agriculture technologies have been included to solve problems of environmental pollution, farm profit, crop yield, resource management, food security, and sustainability issues.

References

[1] Tomich TP, Kilby P, Johnston BF. Transforming agrarian economies: opportunities seized, opportunities missed. Ithaca: Cornell University Press; 2018.

[2] Sui Z, Raubenheimer D, Rangan A. Consumption patterns of meat, poultry, and fish after disaggregation of mixed dishes: secondary analysis of the Australian National Nutrition and Physical Activity Survey 2011–12. BMC Nutrition 2017;3:52. https://doi.org/10.1186/s40795-017-0171-1.

[3] Caprile A, Rossi R. 2021 international year of fruits and vegetables. 2021. https://www.europarl.europa.eu/RegData/etudes/ATAG/2021/689367/EPRS_ATA(2021)689367_EN.pdf.

[4] Kumar A, Nagar S, Anand S. Climate change and existential threats. Glob Clim Change 2021:1–31. https://doi.org/10.1016/B978-0-12-822928-6.00005-8. 9780128229286.

[5] FAO. Nutrition and Food Systems. Rural poverty reduction. FAO Statistics; 2018. http://www.fao.org/in-action/voices-of-the-hungry/fies/en/. [Accessed 16 August 2022].

[6] Mueller N, Gerber J, Johnston M, et al. Closing yield gaps through nutrient and water management. Nature 2012;490:254–7. https://doi.org/10.1038/nature11420.

[7] Zhang N, Wang M, Wang N. Precision agriculture—a worldwide overview. Comput Electron Agric 2002;36(2–3):113–32.

[8] Lee KH, Zhang N. A frequency-response permittivity sensor for simultaneous measurement of multiple soil properties: Part II. Calibration model tests. Transac Asabe 2007;50(6):2327–36.

[9] Vogel E, Donat MG, Alexander LV, Meinshausen M, Ray DK, Karoly D, et al. The effects of climate extremes on global agricultural yields. Environ Res Lett 2019;14(5):054010.

[10] Tesfahunegn GB, Tamene L, Vlek PL. Catchment-scale spatial variability of soil properties and implications on site-specific soil management in northern Ethiopia. Soil Tillage Res 2011;117:124–39.

[11] Pardossi A, Incrocci L, Incrocci G, Malorgio F, Battista P, Bacci L, et al. Root zone sensors for irrigation management in intensive agriculture. Sensors 2009;9:2809–35.

[12] Norrie J, Graham M, Dubé P, Gosselin A. Improvements in automatic irrigation of peat-grown greenhouse tomatoes. Horttecnology 1994;4(2):154–9.

[13] Cormier J, Depardieu C, Letourneau G, Boily C, Gallichand J, Caron J. Tensiometer-based irrigation scheduling and water use efficiency of field-grown strawberries. Agron J 2020;112(4):2581–97.

[14] Caron J, Price JS, Rochefort L. Physical properties of organic soil: adapting mineral soil concepts to horticultural growing media and histosol characterization. Vadose Zone J 2015;14(6):1–14.

[15] Jobin P, Caron J, Bernier PY, Dansereau B. Impact of two hydrophilic acrylic-based polymers on the physical properties of three substrates and the growth of Petunia× hybrida Brilliant Pink. J Am Soc Hortic Sci 2004;129(3):449–57.

[16] Caron J, Xu HL, Bernier PY, Duchesne I, Tardif P. Water availability in three artificial substrates during Prunus× cistena growth: variable threshold values. J Am Soc Hortic Sci 1998;123(5):931–6.

[17] Nguyen TKL, Cho KM, Lee HY, Sim HS, Kim JH, Son KH. Growth, fruit yield, and bioactive compounds of cherry tomato in response to specific white-based full-spectrum supplemental LED lighting. Horticulturae 2022;8:319.

[18] Deram P, Lefsrud M, Orsat V. Supplemental lighting orientation and red-to-blue ratio of light-emitting diodes for greenhouse tomato production. Hortscience 2014;49:448–52.

[19] Hao X, Zheng J, Little C, Khosla S. Hybrid lighting configurations with top HPS lighting and LED inter-lighting and N:K ratios in nutrient feedings affected plant growth, fruit yield, light energy use efficiency in greenhouse tomato production. Louisiana: American Society for Horticultural Science; 2015.

[20] Palmitessa OD, Prinzenberg AE, Kaiser E, Heuvelink E. LED and HPS supplementary light differentially affect gas exchange in tomato leaves. Plants 2021;10:810.

[21] McCree KJ. The action spectrum, absorptance and quantum yield of photosynthesis in crop plants. Agric Meteorol 1972;9:191–216.

[22] Hemming S, de Zwart F, Elings A, Petropoulou A, Righini I. Cherry tomato production in intelligent greenhouses-sensors and AI for control of climate, irrigation, crop yield, and quality. Sensors 2020;20(22):6430. 2020.

[23] Hussain N, Farooque AA, Schumann AW, McKenzie-Gopsill A, Esau T, Abbas F, et al. Design and development of a smart variable rate sprayer using deep learning. Rem Sens 2020;12:4091. https://doi.org/10.3390/rs12244091.

[24] Zaman QU, Esau TJ, Schumann AW, Percival DC, Chang YK, Read SM, et al. Development of prototype automated variable rate sprayer for real-time spot application of agrochemicals in wild blueberry fields. Comput Electron Agric 2011;76(2):175–82.

[25] Farooque AA, Hussain N, Schumann AW, Abbas F, Afzaal H, McKenzie-Gopsill A, et al. Field evaluation of a deep learning-based smart variable-rate sprayer for targeted application of agrochemicals. Smart Agric Technol 2022;3:100073.

[26] Abbas F, Afzaal H, Farooque AA, Tang S. Crop yield prediction through proximal sensing and machine learning algorithms. Agronomy 2020;10(7):1046.

CHAPTER 2

Soil spatial variability and its management with precision agriculture

Humna Khan[1], Travis J. Esau[1], Aitazaz A. Farooque[2,3], Qamar U. Zaman[1], Farhat Abbas[3], Arnold W. Schumann[4]

[1]Department of Engineering, Faculty of Agriculture, Dalhousie University, Truro, NS, Canada; [2]Faculty of Sustainable Design Engineering, University of Prince Edward Island, Charlottetown, PE, Canada; [3]College of Engineering Technology, University of Doha for Science and Technology, Doha, Qatar; [4]Citrus Research and Education Center, Institute of Food and Agricultural Sciences, University of Florida, Gainesville, FL, United States

2.1 Introduction

This chapter explains the concepts of soil and crop variability and its management using precision agriculture (PA). The conventional practices of field management have been indicated as one of the root causes of soil and crop variability. Precision agriculture is supported by various techniques, technologies, and tools including remote sensing, Geographical Information System (GIS), Global Positioning System (GPS), crop management zones, and non-uniform fertilizer application techniques, which help to manage inputs precisely based on soil and crop needs that have been discussed here.

2.1.1 Concepts of variability and precision agriculture

Traditional agriculture applies fertilizers uniformly throughout the fields without considering any variability (spatial and temporal) present in crop parameters and soil properties. Therefore, managing agricultural fields with a uniform application can deteriorate the soil by over-application of agrochemicals in high nutrient zones and under-application in areas having fewer nutrients [1,2].

Poor nutrient management results in the reduction of crop yields especially when the agricultural fields include sloping areas. However, the literature on the mechanism of spatial variability in soil nutrients is still incomplete [3] and a better understanding of this mechanism helps to overcome the adverse impacts of spatial variability on crop production by adopting better soil management practices [4].

Precision agriculture is a goal to understand and practice the development of sustainable agriculture [5]. The variability in the soil must be indicated quantitatively and locally to accomplish the PA. Then profitability and ideal advantages of environmental safety could be achieved by comparing agricultural management practices and land use with the local soil and environmental conditions [3,6]. Different management practices affect the spatial variability of soil, for example, irrigation, fertilization, and formation factors of soil, such as parent materials [7].

Precision agriculture evolved on the estimation of variation present in the field; therefore, the development of management zones is the key to promoting variable management of crop inputs among the different zones [8]. Soil management zones are characterized by homogeneous yield-limiting factors, allowing a site-specific application of agrochemicals for achieving optimum crop yield. Geostatistical methodologies could be applied to evaluate the spatial distribution of soil [9]. Therefore, geostatistics is crucial to address soil variability because it gives the most important information regarding the soil properties this information tells what, when, where, and how much agricultural fertilizers will help in sustaining the productivity of soil and minimizing the costs of production simultaneously with reducing the environmental effect [10]. A deep understanding of the mechanisms of the properties of soil spatial distribution is needed to establish soil management zones and adopt precision farming practices [11,12].

2.2 Soil and crop variability

Spatial and temporal variabilities are the fluctuations in the characteristics of soil and crop as well as the environment, over distance and time, respectively [13]. It is necessary to learn about the spatial variability that exists within the soil properties and crop parameters so that crop production could be precisely managed within a field. So, the management of soil variability [14] present in the field plays a key role in improving crop productivity and reducing environmental risks. Chen et al. [15] reported that spatial and temporal variability exists in the soil properties of agricultural fields. The results suggested that long-term planning can help in reducing the spatial variations which exist in the soil properties. For instance, topographic parameters like slope can be used to apply inputs on a need basis in the sugarcane field which is economically acceptable [16].

Several scholarly sources start debating on fertilizer input decrease and its environmental effects. The application of varying fertilizer rates has caught the attention of academic experts. The literature demonstrates that field variability can be maintained by carefully distributing nutrients based on soil and crop requirements, which contributes to the sustainability of agriculture. Nitrogen is the source of the highest energy input; however, it also contributes to pollution. Morari et al. [17] used varied rates of nitrogen in the wheat crop in the Veneto region of Italy. They applied nitrogen based on NDVI sensors as per crop requirements, resulting in high-quality grain with lower nitrogen inputs. Vatsanidou et al. [18] applied variable amounts of nitrogen depending on the substitution of nutrients depleted by the prior crop. They were able to reduce the applied rate by 43% without impacting the annual production.

Liakos et al. [19] employed uniform and varying rates of fertilizer application in alternate rows of the orchard for 2 years as part of their research related to apples in Greece. He discovered that a significant reduction in nitrogen inputs resulted in a slight loss in production, while the farmer's profit improved. Additionally, he also discovered an improvement in terms of the quality of apples.

2.2.1 Factors affecting soil and crop variability

The quantification of parameters causing spatial variability in crop yield is required to precisely manage the moisture content, organic matter of soil, and other crop inputs in an agricultural field. In a crop production system, the spatiotemporal variabilities present in soil properties of agricultural fields are identified for the assessment of soil health in a site-specific manner. The variability in different properties of soil, for example, chemical, biological, and physical, is the major factor that is responsible for variability in crop yield [20]. Some characteristics of the soil, for example, soil texture, changes slowly over time. However, certain soil characteristics like soil electrical conductivity, holding soil moisture, soil organic matter, or nitrate level change quickly [13].

2.2.2 Monitoring variability in agricultural fields

Advancements in technology have made studying soil variability possible. For example, electromagnetic inductions instrument, such as DualEM-2, determines ground electrical conductivity; time-domain reflectometry (TDR-300) determines soil moisture; soil combustion is used to determine soil organic matter; and sophisticated analytical techniques have become available to analyze soil solutions for their concentrations of a range of soil nutrients including nitrates.

To operate a DualEM-2 sensor, the operator can carry it on his shoulder and walk in the field for data collection (Fig. 2.1A). TDR-300 is manually operated for soil moisture content determination by inserting the sensor's probe into the ground to the depth of interest (Fig. 2.1B). The DualEM-2 and TDR-300 instruments are taken to the nodes of field grids georeferenced with the Real-Time Kinematic assisted GPS (RTK-GPS; Fig. 2.1C). Same grid nodes are used to collect composite soil samples for analyses of organic matter content and nutrient concentration (Fig. 2.1D).

The technologies can collect spatial data for the understanding of changes in soil and crop variabilities [21,22]. These soil characteristics can help to interpret the spatial variations in yield, and to manage the nutrients effectively. The electrical conductivity of soil is defined as the number of total ions present in the nutrient solution extracted from a soil sample. Availability of nutrients in various parts relates to the spatial changes in soil electrical conductivity of an agriculture field, that is, soil fertility in distinctive patches.

FIGURE 2.1

DualEM to record conductivity readings (A), time-domain reflectometer to record moisture content of soil (B), RTK-GPS to locate coordinates of field (C), and composite soil samples for the analyses of organic matter and nutrient concentration (D).

2.3 Precision agriculture

Precision agriculture or site-specific crop management is a farm management idea that is based on the observation, measurement, and response to the inter- and intra-field variations in crops. According to Oliver et al. [6], PA is an approach of advanced informative techniques for providing, processing, and analyzing multi-source data having a high spatiotemporal resolution to make decisions and manage crop production operations. The primary objective of precision farming is to enhance crop yield and minimize adverse environmental impacts, whereas the advantages are: recording the soil and plant physicochemical variables with the help of sensors, that is, electrical conductivity, soil moisture, temperature, radiation, evapotranspiration, etc.

2.3.1 The need for precision agriculture

Over and underapplication of nutrients have cost implications for the farmer, with a waste of material on one hand and a potential loss of yield on the other. Excessive nutrients affect the environment by moving the nitrogen, phosphorus, and pesticides into the surface and groundwater and other areas of land where they are not required. With greater knowledge about soil and crop requirements and crop conditions, fertilizers and pesticides should be applied in a more precise way based on crop and soil requirements. This is a vital role that is embodied in the concept of precision agriculture. For example, remote and proximal sensors are now available that can monitor a crop's requirements and identify weeds and certain crop diseases. The application time of nutrients is also very important for their efficient usage by the crop. The purpose of precision farming is to meet the spatiotemporal demands of the crop with inputs (nutrients, water, and pesticides). A precision agriculture approach has the potential to make modern agriculture more suitable for the environment [6].

The information and communications technologies are utilized in the PA technique to manage the spatiotemporal variations in the agricultural fields. According to another definition, while minimizing the potential environmental pollution, the utilization of modern technologies to enhance crop production is also known as precision agriculture [23]. It can help farmers in the judicious use of their resources, as it gives them knowledge about the use of accurate and advanced inputs that are cost-effective and environmentally friendly. A traceability system can be used for it to record all the site-specific activities [24]. To some, PA is about making decisions for changing agriculture and decreasing agricultural effects on the environment, while to others, it is about prescription maps, grid sampling, application of variable rate inputs, remotely controlled sensors, and vehicles. PA creates site-specific recommendations that use information technologies that account for natural and management-induced changes. However, these both have almost the same advantages that use the information to improve the services produced by the agricultural fields. In the decision process, it is key to point out that PA techniques do not put back the farmer's contribution. The main goals behind PA include upgraded decisions on management, enhanced yields, and lower agricultural effects on the environment [13]. Precision agriculture is a recurring process of different steps that include data collection, data analysis, analysis of results-based crop management decisions, and at the end evaluate the cropping system decisions, this cycle goes on for a later time (Fig. 2.2).

2.3.2 Applications of precision agriculture technology

Precision agriculture technology is applied to develop a decision support system for completing farm management with optimized inputs and outputs of the different systems at regular intervals. It involves data collection for determining the state of the variability of field properties. The data collection can involve soil, water, and/or sampling with manual methods, and the storage and analysis of data in a geographic

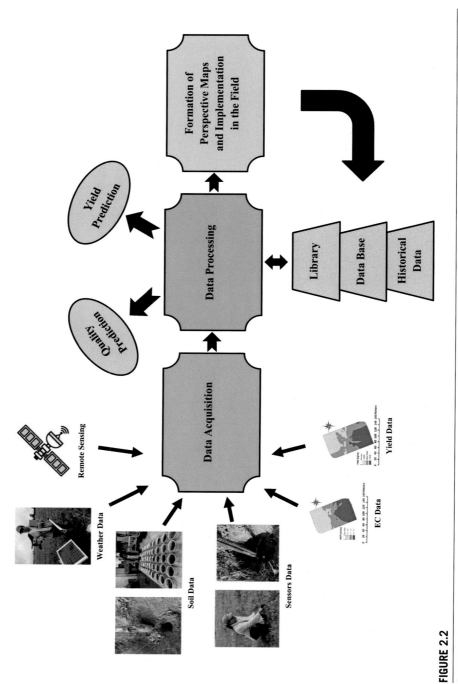

FIGURE 2.2

Precision agriculture system.

information system (GIS). Remote sensing is also used to collect field data for the precision management of crop inputs. It can also be done with an array of multiple sensors in the field.

2.3.2.1 Data collection
Diverse types of soil and crop data can be collected during the growing season to help the growers in decision-making. For example, soil physical and chemical parameters, yield monitoring data, topography, sensors data, and crop scouting data for crop growth and diseases. All the data must be imported into a GIS database with georeferencing using GPS technology. There are different levels of accuracy in GPS technology. The accuracy of most GPS is a few meters. The Differential Global Positioning System has a higher resolution, while RTK-GPS has 1−2 cm accuracy [25].

2.3.2.2 Soil sampling and analysis
Soil is the substrate where crops are grown. It influences various crop parameters, and crop quality and yield. Soil has also been affected by most of the cropping activities like fertilization, tillage, compaction, and so on. From the beginning of precision agriculture, the soil has been analyzed for its physical and chemical properties. Initially, a grid sampling strategy was used. In this method, the field boundary is marked and the whole field is divided into small grids from where samples are collected. The grid size varies according to the objective of the research and the field survey is done before the experiment. Samples collected from different portions of the grid are assorted, integrated, and examined for their physical and chemical properties, for example, cation-exchange capacity, soil texture, pH, organic matter, nutrient content, and so on. Prescription maps for all the soil properties are developed and can be utilized for the implementation of agrochemicals. Fountas et al. [24] used the grid soil sampling and analysis in an olive orchard and developed the soil prescription maps and calculated the required amount of phosphorus and potassium fertilization.

Aggelopoulou et al. [26] performed soil analysis in a dense grid. They did not find consistent relationships between soil properties and yield. Then they recommended delineating prescription maps for the application of fertilizers considering the apple yield and the nutrients removed. Best et al. [27] analyzed 10 samples per hectare and even then found low correlation coefficients among soil variables and yield properties. They proposed that it is better to correlate yield parameters and Apparent Electrical Conductivity (ECa) maps constructed with the data obtained from electromagnetic induction meters, such as DualEM-2. Results revealed that the areas of the fields having high electrical conductivity have high yields and areas having low electrical conductivity have low yields.

2.3.2.3 Precision farming tools to determine the variability
Applying inputs in the right amount at right time is essential to optimize profitability and sustainability without affecting the environment. This can be accomplished

through precision farming. It can be made possible by using the best management practices for minimizing the yield gap by improving food production. It can also assist in maintaining the utilization of natural deposits at an ecofriendly level. Precision farming is a comprehensive technique to increase the productivity of crops using space-based technology and computerized information for assessing and managing the spatiotemporal variations of inputs and resources, for example, chemicals, nutrients, seeds, etc. The application of remote sensing, GPS, and GIS shows a great promise to promote and adopt precision farming to manage spatiotemporal variability at high resolution [28].

2.3.2.4 Geographic information system

A geographic information system (GIS) is a computer-based software, which uses feature characteristics and position data for the development of maps [29]. The main function of GIS in agriculture is to store the instructions layers, such as sensors data, crop scouting data, soil nutrient data, yield, and soil survey maps [30]. Georeferenced data can be presented in the GIS, adding a visible point of view for the explanation. Besides electronic data processing and presentation, the GIS can also be used to estimate, and present alternate management practices by integrating and utilizing data layers to produce an analysis of management plans.

2.3.2.5 Remote sensing

Remote sensing is used to collect field data without physically contacting the target, that is, soil or plant. An electromagnetic wave can be reflects, absorbs, or passes through when falling on an object. Helpful data can be obtained from the agricultural fields and objects by calculating these radiations. It is a very helpful technology for determining soil and crop variability within the field as it provides a wide range of parameter data. All perceivable objects present in the field, that is, soil and plants, can be sensed remotely, based on the different reflections of sunlight from them. Light (sun or some artificial) reflectance has been used as vegetation indices in variability studies. Remote sensing is mostly used in calculating the Normalized Difference Vegetation Index (NDVI). The NDVI is an impression of a plant's health, and it also correlates with the quality of the crop and yield. It is the most used application for the regulation of nitrogen application. The greener plants have high NDVI values which indicate the efficient amount of available nitrogen in them and as a result; these plants require a small quantity of fertilizer as compared to less green plants (lower NDVI) [31]. The NDVI can be calculated as.

$$NDVI = \frac{rNIR - rRED}{rNIR + rRED}$$

where, rNIR: reflection in the Near-Infrared spectrum, rRED: reflection in the infrared spectrum.

2.3.3 Data analysis and delineation of management zones

A sampling of soil and then its analysis is a laborious and expensive task. This can be justified for the sake of research, but it is not sustainable in most commercial applications. The other option is to make the management zones based on electrical conductivity (ECa) or yield mapping and direct the soil sampling of each zone. This option not only decreases the cost and number of samples but also provides a good presentation of the field for crop management. Ferguson et al. [32] used directed sampling, elevation maps, and ECa-based delineated management zones in a vineyard. He reported that nine soil samples were enough to characterize the soil and he also reported that delineation of management zones is a good decision for the farmers to control their management practices. Khan et al. [33] reported that the farmers can adopt nutrient management practices based on the variability in ECa, soil organic matter, soil moisture content, and NDVI because these parameters were easily measured through sensors. It has also been evaluated by Khan et al. [34] that these parameters have a significant correlation with the potato having r ranges from 0.58 to 0.84.

The management zones are defined by dividing the field into different parts. Some field areas have different responses, while others may show similar behaviors [35]. The use of management zones is the most favored method to manage spatial variability within the fields. These zones are field subdivisions that can also be used to direct the variable-rate fertilization [32]. There are various ways for delineating management zones, for example, soil mapping using GIS [36], data collection through proximal sensing of fields, and analysis using deep learning algorithms [37]. Addressing soil and crop variability issues needs the joint development of information database and management skills [30]. Pertinent methods must be used or developed for the examination of data for any missing values and normality issues. The preliminary descriptive statistics give an idea of the values, their distribution, and range. Geostatistics is a probabilistic method of spatial interpolation. Correlations between different soil and crop variables can be conducted to estimate their relationships and the nature of dependence upon one another. The final formation of the maps is made based on the error's estimation at nonsampled points, spatial variability structured by variogram of the sampled data, and an interpolation method (kriging). This type of information that is collected for several properties and consecutive years creates new and fascinating opportunities in agronomic crop analysis and management [38].

Geostatistical methods such as kriging can be used for interpolation between sampling points for the given spatial dependence of the values. Maps can be delineated, and the soil variability and crop parameters can be indicated within the whole field. Kitchen et al. [35] tried to develop geostatistical methods such as kriging to develop management zones based on elevation, ECa, and yield maps with the help of MZA software. Aggelopoulou et al. [39] applied a multivariate technique to delineate the management zones in apples based on soil, crop yield, and quality data. Khan et al. [40] delineated the management zones in potato fields based on electrical conductivity and yield maps using ArcGIS. The data analysis strives for defining the field's parts with common features which can be managed individually. Delineation of management zones should form similar field parts where inputs and other practices can be implemented typically. The management zones should be enough to allow agricultural practices such as using variable rate application of agrochemicals [31].

2.3.4 Variable-rate application technology

The technology to apply nutrients on a need basis is popular among growers to manage spatial variability. Variable-rate identifies us that the suitable number of fertilizers will be spread at the right time in a precise manner, which will reduce inputs, costs, and unfavorable environmental impacts and improve crop yield and quality. For applying variable rates, there are two approaches. The first, referred to as map-based, depends upon historical information (past or current year). Process control technologies use data from GIS-based prescription maps to manage inputs by controlling processes like fertilizer selection and application, seeding rates, and herbicide spreading.

The second method is known as the sensor-based method, which uses sensing technologies that can alter the application rate of different chemicals on the go. The sensors reveal the characteristics of soil and crop and modify the application tool. Based on the delineation of management zones and past information, prescription maps can be developed to define the demands of various parts of the field. Various machines have been designed to adjust the application rate of agrochemicals according to the developed maps of seeding, fertilization, manure, and irrigation applications managed site-specifically.

A comprehensive literature search by Wu et al. [41] revealed that precise nutrient management enhances crop yields by 8%–150% compared with conventional practices, increases in water-use efficiency, and the economic returns to farmers while improving grain quality and soil health, and sustainability.

2.3.4.1 Prescription maps or management zones

Prescription maps or management zones (MZs) can be produced from various properties of the field and the crop [31]. For example, the management zones produced by Khan et al. [40] for potatoes are based on electrical conductivity and yield maps using ArcGIS which helped growers in decision-making (Fig. 2.3). A greater value of conductivity (HCP) and crop yield was noticed in the southeast and southwest areas of the field, respectively. Hence there developed an excellent MZ in the southwest part which means that it was the high-yielding area of the field. Similarly, the other field had greater values of the HCP and yield in the northwest part, that is why it was the highly productive area of the field which has been shown by excellent MZ. Thus, following such techniques, for example, the development of MZs based on soil and crop needs improved soil fertility that also helps in increasing the tuber yield.

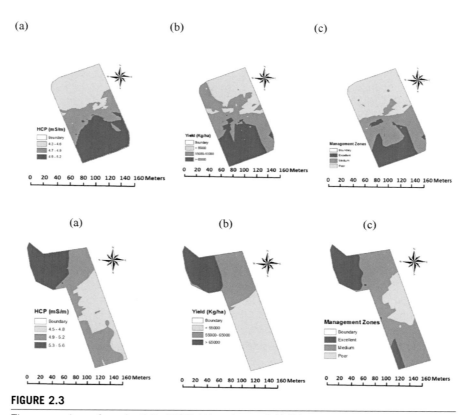

FIGURE 2.3

The comparison of conductivity (HCP) (A) and crop yield (B) were mapped in MZs (C) for the fields of PEI representing the different levels of productivity that required the nutrients based on crop and soil needs across the fields.

Adapted from Khan H, Acharya B, Farooque AA, Abbas F, Zaman QU, Esau T. Soil and crop variability induced management zones to optimize potato tuber yield. Appl Eng Agric 2020;36(4):499–510.

2.3.5 Precision irrigation management to manage spatial variability in soil water content

The water content in the soil is one of the many variables that are impacted by soil spatial variability. To prevent any area of the field from being underirrigated, traditional agricultural techniques modify the irrigation rate for the droughty portions of the agricultural fields. Due to this technique, many other areas of the field use water inefficiently [42]. The highest quantity of water that crops can access in the root zone is determined by the soil's available water-holding capacity [43]. As a result, precise subfield area delineation, such as the delineation of site-specific management zones [44,45], is essential for precision irrigation control [46]. Precision farming relies significantly on soil hydraulic features that vary within a field when defining site-specific management zones. Thanks to the variable-rate irrigation application systems that have made the application of irrigation water at variable rates.

Measuring ECa serves as the foundation for variable-rate irrigation systems. However, depending on the field being researched, different soil characteristics have a greater impact on ECa. Additionally, studies have demonstrated that many soil ECa regulating variables might be present in a single study region [47]. For water management employing precision irrigation systems, a good description of spatial variability in the moisture content present in the soil is essential [42].

2.3.6 Precision agriculture using machine learning

Precision agriculture provides farmers with a variety of techniques designed to achieve optimal outputs with specified inputs. Smart sensors, satellite pictures, actuators, drones, and robots are some of the most important technological advancements that have aided the agriculture business through machine learning. These elements perform a significant character in collecting real-time data and making decisions without a workforce. Artificial intelligence (AI), which is the automation of intelligent behavior, is beneficial for humans in many aspects of life, and our environment and planet [48]. Prediction of soil conditions is the topmost important stage that influences crop selection and yield, land preparation, seed selection, and fertilizer selection. The soil characteristics have a direct relationship with the geographical and climatic circumstances of the study area and are therefore a significant issue to consider. The soil characteristics prediction consists mostly of predicting soil surface, humidity and nutrients, and meteorological conditions during the lifespan of the crop [49]. The crop economy is based on the soil's available nutrients. Electric and electromagnetic sensors provide a substantial measure of data related to soil nutrients [50]. The optimal crop for a piece of land is determined by producers based on nutritional content. The prediction of agricultural yields and methods for maximizing productivity is essential knowledge for any grower. pH value, soil type and quality, and meteorological conditions: temperature, humidity, daylight hours, rainfall, fertilizer, and harvesting schedules are some of the aspects that have a remarkable impact on crop yield prediction.

When properly applied to a system, machine learning models perform as feedforward control. With the assistance of accurate ML models, we can estimate the variables that will influence crop yield. Therefore, corrective action can be conducted before any agricultural production abnormalities [51]. Disease-causing fungus, microbes, and bacteria obtain their energy from the plants they inhabit, which impacts crop productivity. If it is not found on time, it might result in a substantial economic loss for farmers. A significant amount of a farmer's financial strain comes in the form of pesticides, which are used to eliminate illnesses and restore crop functionality. The excessive use of pesticides impacts the agricultural land's water and soil cycles and harms the environment. Using an AI system that is properly designed over the crop growth period not only reduces the danger of crop disease and the economic impact. It also reduces the negative environmental impact of nonsystematic farming [48].

Several descriptive and analytical yield modeling approaches have been employed for various crops [52,53]. These approaches need an immense understanding of the crop and soil conditions, making their implementation in various places challenging. Numerous remote sensing techniques based on satellites are also used in the prediction of crop yield [54,55]; however, these techniques are not able to offer sufficient spatial information on farmlands for their specific crop management to improve fertilizers for crop growth. Latest improvements in machine learning (ML) and data-driven modeling have acquired great demand in this field, allowing researchers to solve and comprehend complex interactions. Different ML approaches have demonstrated the potential for use in crop forecasting [56,57]. In recent years, the advancement of machine learning has the potential to improve farmers' decision-making [58]. Machine learning, a branch of AI, is an effective method that can aid growers in decision-making.

2.3.7 Role of precision agriculture in climate change

One of the worst effects of the evolution of the human race is climate change. Agriculture is regarded as one of the main causes of climate change among the several ones. According to Olsson et al. [59] the agriculture sector, which includes forestry and other land uses, is the second greatest contributor of greenhouse gases (GHG) after the electricity and heat generation sectors, contributing 24% of the total global GHG emissions. The three main GHGs that affect the climate are CO_2, CH_4, and N_2O. The agriculture industry produces large quantities of each of these three gases. Due to an imbalance in the C and N cycle and poor N fertilization management, the agricultural sector is one of the main contributors to rising GHG emissions. GHG emissions can be reduced by improving the timing of fertilizer N application thanks to PA, which enables timely management and application of N fertilizers. This eliminates the addition of excess N to the field. By implementing variable rate N application, Bates et al. [60] predicted a 5% reduction in GHG emission from the baseline GHG emission rate relevant to the production of N fertilizer. Variable rate N application can boost wheat production by 1%−10% while also saving 4%−37% of the total N fertilizer used [61] and 1.25% of the N_2O emissions from the conserved N inputs [62]. When compared to traditional agricultural practices, the adoption of technology like Green Seeker, which is widely utilized in PA, can reduce N use by 21% and save up to 27 kg/N/ha of maize crops. This results in a 5% greater yield. In poor yield areas, variable rate N application could cut N_2O emissions by 34% [63].

In addition to variable rate fertilizer management, variable rate irrigation is another method that lowers GHG emissions. The main determinants of the release of N_2O from the soil are the availability of soil N and the soil's redox potential. Numerous studies have demonstrated that irrigation increases N_2O release significantly compared to nonirrigation [64,65]. When compared to nonirrigated environments, N_2O emissions have been found to rise generally by 50%−140% under irrigated conditions [66]. This emphasizes how crucial it is to manage both N and water in the field to maximize use, which can lower GHG output. While crop demand-driven moisture application would allow maximum use of the applied water and maintain ideal soil moisture without developing excess moisture conditions favoring denitrification, site-specific N management would ensure increased crop yield and maximize the N use efficiency of the crop.

In addition to fertilizer and water management, optimizing the rate of any input application, pesticides including insecticides, herbicides, fungicides, etc., will all lead to reduced emission of GHG through a direct reduction in consumption and ensuring production at the manufacturing level as well as indirectly through increased yield, decreased soil disturbances, and decreased farm operations.

2.3.8 Sustainable agriculture: the ultimate goal

Environmental quality and sustainability can be enhanced by incorporating spatial variability into fertilizer management practices. Using spatial information to better coordinate the crop requirements with nutrient availability will reduce the amount of fertilizer left in the field which could harm the environment through different degradation processes. Assessing the sources and impact of spatial characteristics (e.g., soil type, cover and thickness, water- and nutrient-holding capacity, slope) can assist farmers and planners in selecting the best management strategies for each farm that promote the long-term sustainability of the cropping system.

Sustainability is the combination of economy, society, and ecology. It has been suggested by definitions that sustainable farming should provide high-quality food to a growing human population and ensure food security and stability. It also helps farm owners to protect their health, while also trying to make the foremost use of supplies and preserving them for future generations and protecting the environment and saving the planet from agriculture's negative effects [31].

2.4 Conclusion

Spatial variability exists in the soil properties and crop parameters. When the nutrients are applied uniformly in the fields without considering the spatial and temporal variations, it results in wastage of inputs, as well as the excess amount of fertilizers badly affects the soil and environment. Advanced PA technologies help in the precise management of inputs based on soil and crop need. With the use of various approaches and cutting-edge sensors, PA calculates the variability present in the crop and field and uses the results to effectively meet crop needs. The technology known as variable rate input application allows for the possibility of adjusting inputs to requirements, resulting in decreased inputs and increased yields, improved resource usage, and decreased negative environmental effects. Additionally, PA increases the farms' productivity and profitability. These are the elements that contribute to increased agricultural sustainability. However, adoption is still lagging behind expectations, particularly in many areas with small farms, and their potential advantages should be further investigated.

References

[1] Ferguson RB, Hergert GW, Schepers JS, Gotway CA, Cahoon JE, Peterson TA. Site-specific nitrogen management of irrigated maize: yield and soil residual nitrate effects. Soil Sci Soc Am J 2002;66(2):544−53.

[2] Abd-Elmabod SK, Alí RR, Anaya Romero M, Rosa DD. Evaluating soil contamination risks by using MicroLEIS DSS in El-Fayoum Nile province, Egypt. In: Institute of electrical and electronics engineers. Proceedings of 2010 2nd international conference on chemical, biological and environmental engineering; 2010.

[3] Mansour HA, Abd-Elmabod SK, Engel BA. Adaptation of modeling to the irrigation system and water management for corn growth and yield. Plant Archives 2019;19: 644−51.

[4] Ge F, Zhang J, Su Z, Nie X. Response of changes in soil nutrients to soil erosion on a purple soil of cultivated sloping land. Acta Ecol Sin 2007;27(2):459−63.

[5] Far ST, Rezaei-Moghaddam K. Impacts of the precision agricultural technologies in Iran: an analysis experts' perception & their determinants. Inform Process Agric 2018;5(1):173−84.

[6] Oliver MA, Bishop TF, Marchant BP, editors. Precision agriculture for sustainability and environmental protection. Abingdon: Routledge; 2013. p. 20.

[7] Davatgar N, Neishabouri MR, Sepaskhah AR. Delineation of site-specific nutrient management zones for a paddy cultivated area based on soil fertility using fuzzy clustering. Geoderma 2012;173:111−8.

[8] Castrignanò A, Buttafuoco G, Quarto R, Parisi D, Rossel RV, Terribile F, et al. A geostatistical sensor data fusion approach for delineating homogeneous management zones in Precision Agriculture. Catena 2018;167:293−304.

[9] Mueller TG, Hartsock NJ, Stombaugh TS, Shearer SA, Cornelius PL, Barnhisel RI. Soil electrical conductivity map variability in limestone soils overlain by loess. Agron J 2003;95(3):496−507.

[10] Shaddad SM. Geostatistics and proximal soil sensing for sustainable agriculture. Sustain Agric Environ Egypt: Part I 2018:255−71.

[11] Brevik EC, Calzolari C, Miller BA, Pereira P, Kabala C, Baumgarten A, et al. Soil mapping, classification, and pedologic modeling: history and future directions. Geoderma 2016;264:256−74.

[12] Abd-Elmabod SK, Fitch AC, Zhang Z, Ali RR, Jones L. Rapid urbanisation threatens fertile agricultural land and soil carbon in the Nile delta. J Environ Manag 2019;252: 109668.

[13] Kent Shannon D, Clay DE, Sudduth KA. An introduction to precision agriculture. Precis Agric Basics 2018;11:1−2.

[14] Ziadi N, Cambouris AN, Nyiraneza J, Nolin MC. Across a landscape, soil texture controls the optimum rate of N fertilizer for maize production. Field Crop Res 2013;148: 78−85.

[15] Chen S, Lin B, Li Y, Zhou S. Spatial and temporal changes of soil properties and soil fertility evaluation in a large grain-production area of subtropical plain, China. Geoderma 2020;357:113937.

[16] Sanches GM, Magalhães PS, dos Santos Luciano AC, Camargo LA, Franco HC. Comprehensive assessment of spatial soil variability related to topographic parameters in sugarcane fields. Geoderma 2020;362:114012.

[17] Miao Y, Mulla DJ, Robert PC. Identifying important factors influencing corn yield and grain quality variability using artificial neural networks. Precis Agric 2006;7(2):117—35.
[18] Gonzalez-Sanchez A, Frausto-Solis J, Ojeda-Bustamante W. Predictive ability of machine learning methods for massive crop yield prediction. Spanish J Agric Res 2014;12(2):313—28.
[19] Morari F, Loddo S, Berzaghi P, Ferlito JC, Berti A, Sartori L, et al. Understanding the effects of site-specific fertilization on yield and protein content in durum wheat. In: Precision agriculture'13. Wageningen: Wageningen Academic Publishers; 2013. p. 321—7.
[20] Awal R, Safeeq M, Abbas F, Fares S, Deb SK, Ahmad A, et al. Soil physical properties spatial variability under long-term no-tillage corn. Agronomy 2019;9(11):750.
[21] Serrano JM, Peça JO, da Silva JR, Shaidian S. Mapping soil and pasture variability with an electromagnetic induction sensor. Comput Electron Agric 2010;73(1):7—16.
[22] Machado FC, Montanari R, Shiratsuchi LS, Lovera LH, Lima ED. Spatial dependence of electrical conductivity and chemical properties of the soil by electromagnetic induction. Rev Bras Ciência do Solo 2015;39:1112—20.
[23] Khosla R, Shaver T. Zoning in on nitrogen needs. Colorado State Univ Agron Newslett 2001;21(1):24—6.
[24] Fountas S, Bartzanas T, Bochtis D. Emerging footprint technologies in agriculture, from field to farm gate. Intell Agrifood Chains Networks 2011;25:67—85.
[25] Heraud JA, Lange AF. Agricultural automatic vehicle guidance from horses to GPS: how we got here, and where we are going. In: 2009 Agricultural equipment technology conference, Louisville, KY; 2009. p. 9—12.
[26] Aggelopoulou AD, Bochtis D, Fountas S, Swain KC, Gemtos TA, Nanos GD. Yield prediction in apple orchards based on image processing. Precis Agric 2011;12(3):448—56.
[27] Best S, León L, Claret M. Use of precision viticulture tools to optimize the harvest of high-quality grapes. Proc Fruits Nuts Vegetable Produc Eng TIC (Frutic05) 2005;12:249—58.
[28] Mani PK, Mandal A, Biswas S, Sarkar B, Mitran T, Meena RS. Remote sensing and geographic information system: a tool for precision farming. Geospatial Technol Crops Soils 2021:49—111.
[29] Kitchen NR, Clay SA. Understanding and identifying variability. Precis Agric Basics 2018;05:13—24.
[30] Fulton JP, Port K. Precision agriculture data management. Precis Agric Basics 2018:169—87.
[31] Iakovou E, Bochtis D, Vlachos D, Aidonis D. Supply chain management for sustainable food networks. John Wiley & Sons; 2016. p. 22.
[32] Ferguson RB, Lark RM, Slater GP. Approaches to management zone definition for use of nitrification inhibitors. Soil Sci Soc Am J 2003;67(3):937—47.
[33] Khan H, Acharya B, Farooque AA, Abbas F, Zaman QU, Esau T. Soil and crop variability induced management zones to optimize potato tuber yield. Appl Eng Agric 2020;36(4):499—510.
[34] Khan H, Esau T, Farooque AA, Abbas F, Zaman QU, Barrett R, et al. Identification of significant factors affecting potato tuber yield for precision management of soil nutrients. Appl Eng Agric 2021;37(3):535—45.
[35] Kitchen NR, Sudduth KA, Myers DB, Drummond ST, Hong SY. Delineating productivity zones on claypan soil fields using apparent soil electrical conductivity. Comput Electron Agric 2005;46(1—3):285—308.

[36] Gozdowski D, Stepień M, Samborski S, Dobers ES, Szatyłowicz J, Chormański J. Determination of the most relevant soil properties for the delineation of management zones in production fields. Commun Soil Sci Plant Anal September 25, 2014;45(17):2289—304.

[37] Abbas F, Afzaal H, Farooque AA, Tang S. Crop yield prediction through proximal sensing and machine learning algorithms. Agronomy 2020;10(7):1046.

[38] Arnó Satorra J, Martínez Casasnovas JA, Ribes Dasi M, Rosell Polo JR. Precision viticulture. Research topics, challenges and opportunities in site-specific vineyard management. Spanish J Agric Res 2009;7(4):779—90.

[39] Aggelopooulou K, Castrignanò A, Gemtos T, De Benedetto D. Delineation of management zones in an apple orchard in Greece using a multivariate approach. Comput Electron Agric 2013;90:119—30.

[40] Khan H, Farooque AA, Acharya B, Abbas F, Esau TJ, Zaman QU. Delineation of management zones for site-specific information about soil fertility characteristics through proximal sensing of potato fields. Agronomy 2020;10(12):1854.

[41] Wu W, Ma B. Integrated nutrient management (INM) for sustaining crop productivity and reducing environmental impact: a review. Sci Total Environ 2015;512:415—27.

[42] De Lara A, Khosla R, Longchamps L. Characterizing spatial variability in soil water content for precision irrigation management. Agronomy 2018;8(5):59.

[43] Peters RT, Desta KG, Nelson L. Practical use of soil moisture sensors and their data for irrigation scheduling. 2000. Available online: https://rex.libraries.wsu.edu/esploro/outputs/report/Practical-use-of-soil-moisture-sensors/99900501891201842.

[44] Fleming KL, Westfall DG, Wiens DW, Rothe LE, Cipra JE, Heermann DF. Evaluating farmer developed management zone maps for precision farming. In: Proceedings of the fourth international conference on precision agriculture; 1999. p. 335—43.

[45] Khosla R, Fleming K, Delgado JA, Shaver TM, Westfall DG. Use of site-specific management zones to improve nitrogen management for precision agriculture. J Soil Water Conserv 2002;57(6):513—8.

[46] Hedley CB, Yule IJ, Tuohy MP, Vogeler I. Key performance indicators for simulated variable-rate irrigation of variable soils in humid regions. Trans ASABE 2009;52(5):1575—84.

[47] Tang S, Farooque AA, Bos M, Abbas F. Modelling DUALEM-2 measured soil conductivity as a function of measuring depth to correlate with soil moisture content and potato tuber yield. Precis Agric 2020;21(3):484—502.

[48] Sharma A, Jain A, Gupta P, Chowdary V. Machine learning applications for precision agriculture: a comprehensive review. IEEE Access 2020;9:4843—73.

[49] Parikh SJ, James BR. Soil: the foundation of agriculture. Nat Educ Knowl 2012;3(10):2.

[50] Adamchuk VI, Hummel JW, Morgan MT, Upadhyaya SK. On-the-go soil sensors for precision agriculture. Comput Electron Agric 2004;44(1):71—91.

[51] Bhanumathi S, Vineeth M, Rohit N. Crop yield prediction and efficient use of fertilizers. In: 2019 international conference on communication and signal processing (ICCSP). IEEE; 2019. p. 0769—73.

[52] Cousens R. An empirical model relating crop yield to weed and crop density and a statistical comparison with other models. J Agric Sci 1985;105(3):513—21.

[53] Dourado-Neto D, Teruel DA, Reichardt K, Nielsen DR, Frizzone JA, Bacchi OO. Principles of crop modeling and simulation: I. Uses of mathematical models in agricultural science. Sci Agric 1998;55:46—50.

[54] Doraiswamy PC, Moulin S, Cook PW, Stern A. Crop yield assessment from remote sensing. Photogramm Eng Rem Sens 2003;69(6):665–74.

[55] Prasad AK, Chai L, Singh RP, Kafatos M. Crop yield estimation model for Iowa using remote sensing and surface parameters. Int J Appl Earth Obs Geoinf 2006;8(1):26–33.

[56] Kaul M, Hill RL, Walthall C. Artificial neural networks for corn and soybean yield prediction. Agric Syst 2005;85(1):1–8.

[57] Vatsanidou A, Fountas S, Nanos G, Gemtos T. Variable rate application of nitrogen fertilizer in a commercial pear orchard. From Fork Farm: Int J Am Farm School Thessaloniki 2014;1(1):1–8.

[58] Liakos V, Tagarakis A, Vatsanidou A, Fountas S, Nanos G, Gemtos T. Application of variable rate fertilizer in a commercial apple orchard. Precis Agric 2013;13:675–81.

[59] Olsson L, Opondo M, Tschakert P, Agrawal A, Eriksen S, Ma S, et al. Livelihoods and poverty: climate change 2014: impacts, adaptation, and vulnerability. Part A: global and sectoral aspects. Contribution of working group II to the fifth assessment report of the intergovernmental panel on climate change. In: Climate change 2014: impacts, adaptation, and vulnerability. Part A: global and sectoral aspects; 2014. p. 793–832.

[60] Bates J, Brophy N, Harfoot M, Webb J. Sectoral emission reduction potentials and economic costs for climate change SERPEC-CC. Agriculture: methane and nitrous oxide. 2009.

[61] Arif BT. Variable rate fertilizer application in Turkish wheat agriculture: economic assessment. Afr J Agric Res 2010;5(8):647–52.

[62] Paustian KE, Babcock BA, Hatfield J, Kling CL, Lal RA, McCarl BA, et al. Climate change and greenhouse gas mitigation: challenges and opportunities for agriculture. CAST (Counc Agric Sci Technol) Task Force Rep 2004;15:141.

[63] Sehy U, Ruser R, Munch JC. Nitrous oxide fluxes from maize fields: relationship to yield, site-specific fertilization, and soil conditions. Agric Ecosyst Environ 2003;99(1–3):97–111.

[64] Livesley SJ, Dougherty BJ, Smith AJ, Navaud D, Wylie LJ, Arndt SK. Soil-atmosphere exchange of carbon dioxide, methane and nitrous oxide in urban garden systems: impact of irrigation, fertiliser and mulch. Urban Ecosyst 2010;13(3):273–93.

[65] Liu C, Wang K, Meng S, Zheng X, Zhou Z, Han S, et al. Effects of irrigation, fertilization and crop straw management on nitrous oxide and nitric oxide emissions from a wheat–maize rotation field in northern China. Agric Ecosyst Environ 2011;140(1–2):226–33.

[66] Trost B, Prochnow A, Drastig K, Meyer-Aurich A, Ellmer F, Baumecker M. Irrigation, soil organic carbon and N_2O emissions. Rev. Agron Sustain Dev 2013;33(4):733–49.

CHAPTER 3

Geospatial technologies for the management of pest and disease in crops

Manjeet Singh[1], Aseem Vermaa[1], Vijay Kumar[2]

[1]*Department of Farm Machinery and Power Engineering, Punjab Agricultural University (PAU), Ludhiana, Punjab, India;* [2]*Department of Entomology, Punjab Agricultural University (PAU), Ludhiana, Punjab, India*

3.1 Introduction and objectives

Geospatial Technology (GST) refers to all emerging technologies, such as Remote Sensing (RS), Geographical Information Systems (GIS), and Global Positioning Systems (GPS), that help the user in the collection, analysis, and interpretation of spatial and temporal data. Geospatial technology has been employed in agriculture since the early 1990s. Farmers can now utilize a technology called the Global Navigation Satellite System (GNSS) to identify their agricultural equipment while also analyzing the spatial and temporal variation in soil, relief, and vegetation. This is possible because to the advancement of this system. This technology also improves the quality and accessibility of digital geographic data.

Managing pests and diseases through precision farming is not a new concept. Aerial image analysis for epidemic detection has been practiced since the 1920s. The precision farming for pest management can be used in three ways.

1. Targeted treatments for pest and disease incidence and outbreaks
2. Comprehensive early detection of outbreaks of pests and diseases
3. Early detection of pest and disease outbreaks through sampling

Different technologies are applied at various growth stages of crops. It is possible to detect changes in plant health early enough to prevent the spread of an outbreak, and treatment costs can be reduced by targeting these regions with remote sensing.

Geospatial technologies are particularly useful in the field of disease and pest management, which involves treating only those parts of the field those are at risk of being infested by pests or diseases and resulting severe yield losses. Some parts of a field are implicitly more vulnerable to epidemics/outbreaks than others, regardless of the specific pest or disease in crops. This is because of subtle variations in temperature, precipitation, humidity, and other environmental parameters, especially microenvironment in crops with open or weak canopies. The difficulty is compounded by the fact that these "hotspots" may move around during the year.

Insects and diseases can be detected by aerial photography with optical or thermal cameras. With the help of variable rate application (VRA) technology and remote sensing data that highlight possible trouble sites in the fields, farmers can treat crops just in the sites where a treatment is required. Sprayers with sectional control or individual nozzle level with real-time kinematic (RTK)–based GPS control precisely the targeted applications.

It is widely known that aerial imaging can assist growers to reduce treatment costs by identifying areas likely to be damaged by pests or diseases. However, research is currently ongoing in regard to the early identification or forecasting of these outbreaks. Not every hotspot shown in aerial imaging is brought on by pathogens or pests. Even when they are, it may not always be possible to pinpoint the particular cause of crop stress using only remote sensing data. Therefore, scouting and ground-truthing are still necessary even with the use of remote sensing data.

Unlike thorough early detection, which depends on aerial images of the field being examined, sampling early detection entails taking incredibly comprehensive imagery from various fields and extrapolating results about a broader area from these samples. Automated detection of plant pests, diseases, and general stress can be achieved by analyzing microscopic images of plant parts like stems and leaves. There are, however, three major challenges to this anticipated development. First, the technique is dependent on algorithms that are driven by artificial intelligence, and in order for it to be called reliable, it will require further improvement. Second, getting this method to produce on a massive scale, when millions of acres are the goal, is challenging. Third, there will be a slow adoption of early detection technology because of issues in network connectivity. Hence, keeping in mind all these issues, the major objectives of the chapter are:

- To impart knowledge regarding the overview of scouting and precision in pest management
- To know about various sensing technologies for the detection and monitoring of insect pests and disease
- To develop spatially variable rate technology for the application of chemicals

3.2 Brief review of literature

Although the usage of geospatial data in precision farming is under rigorous research and development, it still promises a huge impact on the way agriculture is practiced. It would be interesting when it will be released with full potential for commercialization in the agricultural fields. However, precise crop management via geospatial technologies faces a number of challenges and possibilities.

According to Mandal et al. [1], Brazil utilized the geostatistics method to generate a geographically differentiated nematode risk map in the field in order to address the issue of a loss in output of monoculture cotton of up to 40% which was brought on by an infestation of *Rotylenchulus reniformis*. In order to assess the risk of insects and diseases and the crop cultural requirements, ground-based weather, plant-stage measurements, and distant imaging were georeferenced in GIS software.

Models for 6 insect pests and 12 crop diseases were calculated for agricultural survey regions and provided daily in georeferenced maps. Remote sensing can also be utilized to evaluate plant nutritional requirements, according to Wojtowicz et al. [2]. Two groups of cucumber plants (*Cucumis sativus*) were studied; one group was inoculated with the bacterial wilt-causing pathogen *Ralstonia solanacearum*, while the other group served as a control. Both groups of cucumber plants were subjected to variations in nutritional content, water content, and light exposure during the presymptomatic stage, all of which can be used in the field. A spectro radiometer was used to determine the cucumber's spectrum reflectance. When a plant is infected with bacteria, viruses, or fungi, there is often a period where the plant shows no outward signs of illness. It takes time for pathogens to become established and develop within plants, or the plant's defense mechanism kicks in. In China, Liu et al. characterized and estimated the fungal disease severity of rice brown spots with hyper-spectral reflectance data [3]. The findings not only show the capability of hyper-spectral remote sensing data in describing plant disease for precision pest management in the real world, but also demonstrate that crop reflectance ratio is a suitable tool for estimating crop disease severity.

In Australia, hyper-spectral sensing was used to identify pests and diseases in vegetable crops by comparing reflectance data for various symptom sets [4]. Another study was also conducted using hyperspectral remote sensing as a tool for the early detection of leaf rust in blueberries. In order to keep track of information gathered from pheromone traps and field scouting, the FAO has created an android app called FAMEWS. Improvements to crop monitoring can be made through studies of remote sensing, image analysis of insects and damage, and automatic counting of trap catches. It will contribute to understanding fall armyworm (*Spodoptera frugiperda*) biology, as well as provide opportunities for forecasting.

Geotagged photos were created using cutting-edge digital imaging technology that included GPS locations and time stamps for use in the control of plant pests and diseases [5]. Using Exchangeable Image File Format (EXIF) metadata, a geocoded image can be easily uploaded to a computer for use with a mapping system, allowing for real-time monitoring or mapping of the spatial distribution of pests and plant diseases. By fusing online mapping with digital image technology, integration mapping software streamlines the mapping process and improves the timeliness and accuracy with which pests and plant diseases may be monitored. The precise locations of areas with positive Plum Pox Virus (PPV) and all PPV-negative prunus block [6] were obtained using GPS technology for use in managing plant pests and diseases. As a major virus, PPV poses a significant risk to the agricultural industry because of the devastating impact it will have on fruit production in terms of both quantity and quality. Meanwhile, GPS was employed to track the locations of diseased and uninfected areas in the fight against Panama disease, which had afflicted bananas in the Middle East [7]. Aguilar et al. used a geolocation method with GPS technology for in situ habitat characterization in managing banana insect pests in Mindanao, Philippines [8]. The region was observed for the presence of insect pests at monthly intervals for 6 months. Then, each location was compared to analyze the presence and incidence of the insect pests infesting the crops. In Malaysia, data collection using handheld GPS enabled the identification of the hotspots infested with disease and pest.

3.2.1 Case studies

Based on weather conditions, a web-based decision support system for potato late blight disease was developed [9]. The study's main goal was to create a basic computer-based forecasting model and post it online for potato producers to easily access. Based on relative humidity (RH) and temperature over the previous week, a model was created that estimates the likelihood of late blight scenarios in terms of disease severity values. A total of six automatic weather stations were installed at different locations in potato-growing areas of Punjab (India). These weather stations' hourly data of temperature, relative humidity, leaf wetness, and rainfall were retrieved daily.

The most favorable weather conditions for late blight development are the temperature between 10°C–20°C, relative humidity of >90%, rainy spells, cloudiness, and foggy weather. The risk evaluation model based on the duration of RH > 90% at a range of temperatures was used to calculate disease severity values. Based on information about prevalent weather in a particular location, likely disease severity was calculated. Accordingly, the farmers were issued advisories on whether or not to initiate fungicide applications. This proved very effective for timely controlling late blight and saving money on fungicide sprays.

Another study examined at the application of IoT in an agricultural field setting to keep tabs on environmental factors including temperature, humidity, rainfall, and wind speed in order to foresee an onslaught of whiteflies and implement preventative measures [10]. Prevalent conditions in the field improve the necessities to better serve the aim of prediction. To minimize the likelihood of a whitefly infestation in cotton, these forecasts are fed into a deep learning model during training and evaluation. In terms of suggesting pesticides, the forecast worked out extremely well. Accurately increasing crop productivity is made possible by the use of real-time sensor-based environmental data, such as temperature, humidity, rainfall, and wind speed. There have been significant improvements in preventing whitefly infestations in cotton crop areas after the introduction of the proposed methodology (Fig. 3.1).

Similarly, the Xarvio scouting app—digital farming solutions by BASF are available in English, Hindi, Marathi, Tamil, Telugu, Kannada, and Malayalam, [11]. The app detects and identifies in-field stress simply by taking a photo using a smartphone. Xarvio SCOUTING is designed to support agricultural advisors in their decision-making just by walking over the fields and taking pictures. Scouting supports farmers with many options including weed identification, yellow trap analysis, detecting both species present and population density, disease recognition, nitrogen recommendation, and leaf damage detection. Xarvio SCOUTING is based on deep learning algorithms, hence continuously learning and improving.

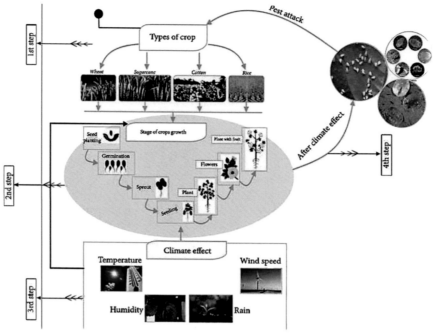

FIGURE 3.1

Model for climate role in the growth of whitefly insect pests.

3.3 Precision in pest management

Both precision agriculture (PA) and plant phenotyping rely heavily on information and technological advancements, and each presents its own unique set of obstacles and requirements for the identification and diagnosis of plant diseases. Precision agriculture is a crop management technique that is based on a field's soil and crop elements' temporal and spatial variability [12]. The goal of this system is to improve management decision-making by creating accurate, real-time maps of crops, soil, and environmental factors. Optical sensing techniques are useful for mapping out fields and pinpointing key disease sites and areas with varying disease severity [13], since the development of plant disease depends on specific environmental conditions and because illnesses often show a heterogeneous distribution in fields. These methods can be combined with sophisticated data analysis tools to create tailored pest management techniques for sustainable crop production. Precise, site-specific, and targeted pesticide treatments result in the possible reduction in pesticide use, which in turn has a lesser economic and ecological impact on agricultural crop production systems [14].

Plant phenotyping evaluates how a genotype looks and performs in different environments, while plant agronomy studies spatial variations within crop stands.

Genotyping and phenotyping must be used effectively in disease-resistance breeding to assess host-pathogen interactions and the vulnerability of the breeding material. Plant phenotyping is labor-intensive, expensive, and time-consuming. Phenotyping is a term that has recently been used frequently to refer to noninvasive imaging and sensor-based characterization of morphological, physiological, and biochemical plant features [15,16].

Plant phenotyping and precision agriculture each have their own set of requirements and challenges when it comes to identifying plant diseases. New strategies must be developed and integrated into conventional monitoring and rating systems in order to achieve an objective and reliable automated detection and diagnosis of plant diseases. Optical sensors have significant promise for noninvasive diseases diagnosis and detection. To assist in diagnosis and plant disease detection, a significant number of imaging and noninvasive sensors are available. Precision agriculture and plant phenotyping now have more prospects due to the advancement of sensor and information technologies as well as the growth of geographic information systems (Fig. 3.2). These sensors can be used on many sizes, from individual cells to entire ecosystems, for plant phenotyping and precision agriculture applications. Several platforms can be employed, depending on the scale, and as a result, different plant metrics can be seen.

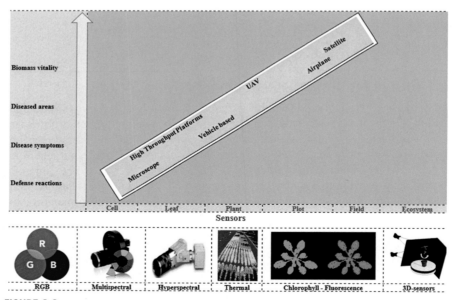

FIGURE 3.2

Overview of existing sensor technologies considered for the automated sensing and identification of host—plant interactions by Oerke et al. [17].

3.4 Overview of scouting

Accurate pest identification, timing/schedule of scouting, location, level of infestation, correct management, and posttreatment follow-up are the major components of scouting for site-specific pest management. The type of pest determines the scouting schedule, frequency, and protocols [18]. Scouting entails spotting bug or disease specimens or signs of plant damage visually [19]. Scouting for a particular pest or disease requires very specific methods which can vary depending upon the pests, stage of crop, within plant sampling or in the soil, as well as economic threshold level. Insects/pests on plants and in soil can be identified and their populations estimated by a number of different scouting techniques used in integrated pest management (IPM). In IPM, which involves precise pest identification, scouting is the initial step.

There is no one method suitable for scouting insects and diseases. The precision of bug counts can be improved by using a combination of scouting methods. Scouting efforts should always be documented for reference. During critical crop stages, examine more plants for accurate quantification of the pests. Beneficial insect populations such as lady beetles, lacewings, spiders, and predatory bugs should also be recorded which can help in environmental conservation efforts.

Aerial, drone, or satellite images can be used to gather information for target scouting. It is crucial to remember that remote sensing is a technique that aids in the detection of irregular patterns in the field, but ground-truth data must be collected before management can be defined. Changes in the species, size, and scope of the problem must be monitored when managing pests. Due to changing environmental factors and the possibility of an epidemic (outbreak), this calls for routine field scouting. In order to track changes in the scope and extent of problem regions, it is essential to scout fields multiple times during the course of the growing season. This is because the incidence of pests in the field often differs in time and space due to variations in environmental conditions and the epidemic potential of diverse pests.

3.4.1 Pest identification

The major bottleneck in field crop production is attack of insect pests and diseases and presence of weeds which causes significant reduction in crop yield. The effective management of these pests depends upon their correct identification. Incorrect identification can lead to improper control tactics that cost time and money, for example, immature of beetle (grub) may look like a caterpillar or worm. The traditional approach to identify pests is very human-resource driven and prone to inefficiencies, and difficult to scale and sustain.

Various tools and resources are available worldwide for pest identification which include different wildlife pest identification tools (i) Western IPM Center, UC Davis (ii) Plant Health Identification Aids and Services U.S. Department of Agriculture (USDA) (iii) Bug Guide, Identify US, Pest Fact Sheets—Clemson

University, etc. [20]. The top five apps on iPhone are Picture Insect, Insect Identification, Insect id: Bug Verifier, Insect identifier Bug Finder, and Smart Identifier for the identification of bugs and insects. With the help of these applications, identifying insects on your iPhone and learning more about them is quick and easy. Picture Insect has more than 1000 species of insect in its database with an identification accuracy rate of 95%. Insect identification is a simpler app than Picture Insect which allows you to either take a photo directly or import a photo from the camera. The app will try to match the photo to an insect from its database. The insect id: Bug Verifier app is easy to navigate. The Smart Identifier helps to identify plants as well as insects from its database [21,22].

Among the various technologies used to identify insect pests, AgSmartic's indigenous technology called Pestlytics is an IoT and ML-based pest detection and monitoring system, to identify pests and provide advisories related to pest attack to the stakeholders right at the appearance of first pests and even before the initiation of the egg laying. Through this intervention, pests can be detected and predictions can be made for effective management of the pests. Prediction of pests and advisory system is built based on the analytics of data gathered from various sources, sensors, crop models, weather, etc., in various geographical locations. The advisory is notified to the extension functionaries of central/state governments, NGOs, and farmers in the catchment area about pest densities, pest forecast of significant size, probability or extent of crop damage. These reports will aid in decision-making for adopting sustainable plant protection strategies by minimizing the dependence on pesticides and encouraging residue-free safe produce [23]. Similarly, Xarvio scouting app—Digital farming solutions by BASF detect and identify in-field stress simply by taking a photo with a smartphone [11].

3.4.2 Insect/pest assessment and control

The economic threshold levels (ETLs) of important insects/pests have been established in order to take appropriate control measures when the pest population reaches a level that may cause economic damage [24]. The knowledge of ETLs aid in crop loss reduction, need-based pesticide application, and increased profit.

To find species and investigate population distribution and movements (migrations in the environment), researchers may employ light traps, yellow pans with water, or sticky-colored traps. Sex pheromone-—baited traps have recently been developed and are currently being tested for various species. Ecofriendly pest control, such as mating disruption, is being explored for large-scale pest management in cotton [25]. A synthetic pheromone, a volatile organic chemical, has been developed to mimic the species-specific sex pheromone released into the air by female insect.

3.5 Integrated pest management (IPM)

By combining the best practices of cultural, biological, mechanical, chemical, and biotechnological pest management, IPM can be used to protect the environment while also boosting productivity and saving money [19]. The effectiveness of the various methods employed is dependent on the accuracy of the pest population monitoring method used. Basic principles of IPM involves monitoring, identifying pests, determining threshold levels, evaluating pest management strategies, selecting pest management strategies, evaluating the results, and making adjustments. IPM includes the selection of the most effective pest management strategy that will do the least harm to people, nontarget organisms, and the environment.

The spatial distribution of insect and diseases and its use in IPM via Geographic Information System (GIS_help in making maps with different formats of points, lines or polygons, and overlap layers [26]. This kind of mapping has been done to indicate higher and lower density of *Grapholita molesta* for effective and accurate management decisions based on insect density. These kinds of techniques are more useful in area-wide pest management [26]. For accurate pest predictions, it's useful to be able to look at historical data, which is made easier by using a GIS map. As a result, GIS has given scientists and policymakers a new tool for tracking and predicting insect pests in crops using a range of geospatial maps. It can significantly reduce the massive crop losses caused by insect pests in agriculture. The spatial complexity of the biotic and abiotic characteristics of a field and its crop, as well as information on the disease and pest populations present, can be fully understood with the help of Geospatial Technologies in an IPM context through remote mapping or spatial modeling.

GIS can be used to assess the spatial pattern of pest infestations, leading to more cost-effective and efficient containment and management of the affected area. Information on insect populations and biotic interactions is crucial for IPM-based control, since it allows for the creation of action plans according to the biology of different insect pests and natural enemies. Kriging, according to the research, is a robust method of spatial interpolation because it employs sophisticated mathematical formulas to infer the values of unknowns from the values of knows.

3.6 Sensing systems for detection and monitoring of pests and disease

Plant protection might be considerably aided by the widespread use of noncontact, effective, and economical methods for spotting and tracking plant diseases and pests [27]. Different sensing systems can be organized into different categories based on their capabilities and level of development in identifying and tracking plant diseases and pests, including:

1. Synthetic aperture radar (SAR)
2. Light detection and ranging (LIDAR) systems
3. Visible and near-infrared (VIS-NIR) based spectral sensors
4. Fluorescence and thermal sensors

A number of remote sensing features are employed in the detection and monitoring processes based on the data collected from these remote sensing devices and sensitivity analysis. These include:

1. Optical, fluorescence, and thermal parameters
2. Image-based landscape features
3. Features associated with habitat suitability

The most widely used remote sensing systems for keeping an eye on plant diseases and pests are based on sensors that detect light in the visible and near-infrared spectrum. The band reflectance feature of VIS-NIR is frequently used for monitoring [27]. Researchers have shown that certain plant diseases and pests are most detectable in the green, red, and NIR spectral ranges. The red-edge peak shifted noticeably to the blue as measured by the first derivative value in rice plants with rice leaf folder infestation [28]. Vegetation indices (VIs) are employed extensively in plant disease detection and pest monitoring, besides reflectance amplitude and converted spectral bands. Spectral responses that are both robust and unambiguous can be triggered by changes in plant pigments including chlorophyll, pigments ratio (Car/Chl), carotenoid, anthocyanin, and the xanthophyll cycle [29,30]. Some VIs, such as the Triangular VI, Normalized Difference Vegetation Index (NDVI), Green NDVI, and Optimized Soil-Adjusted Vegetation Index (OSAVI), are susceptible to a wide range of diseases/pests [29,31,32] because they are affected by many plant characteristics. Xu et al. [33] utilized a water band index (IWB) to identify leaf miner infestation in tomatoes because of the correlation between plant wilting and the presence of a disease or pest. The water index (WI) was discovered to be susceptible to leaf roll disease of grapevine by Naidu et al. [34]. Moreover, new VIs were designed to specifically identify various diseases or pests. Apan et al. [35] created a disease-water stress index (DSWI) for identifying sugarcane orange rust. Aphid infestations can be identified with the help of a proposed aphid index (AI) [36]. Similarly, a leaf hopper index (LHI) depending on detected sensitivity bands was recommended for use in cotton leaf hopper identification [37].

With the advent of fluorescence and thermal sensors for disease and pest detection in plants, a few factors were developed to link the signals with symptoms of infection. Particular investigations have used a ratio of fluorescence (e.g., F686/F740) amplitude at fluorescence peaks to accomplish presymptomatic identification for some diseases, using data from consistent measurements of fluorescence [38,39]. A number of fluorescence parameters, including the maximum quantum efficiency of photosystem II (PSII), the maximum efficiency of PSII photochemistry in light-adapted material (Fv'/Fm'), primary photochemistry (Fv/Fm), nonphotochemical quenching (NPQ), and the effective quantum yield of photosystem II (PSII) have been adopted for use in the detection of plant diseases and pests based on the saturation pulse fluorescence analysis [40]. Features acquired from thermal infrared sensors are less complex for identifying plant diseases and pests than spectral features and fluorescence features. The temperature of the canopy or the leaves is the most significant indicator. It was discovered that in a greenhouse setting, the temperature gradient between the leaf and the air could be used to detect an infection with downy mildew in grapevines before any symptoms appeared [41]. It was also shown that a temperature map of the canopy acquired with an IR camera can reveal the presence of cotton root rot at an early infection stage [42].

Optical, thermal, and fluorescent data can all provide useful context for understanding infection symptoms; by combining these, we can hopefully improve our ability to spot signs of strain [43,44]. For instance, optical, fluorescent, and thermal features were used to detect *Verticillium* wilt of olive at an early stage and differentiate between different infection densities.

Using texture or landscape characteristics, the image-based approach enables precise mapping of diseased areas and estimates of severity. Texture features (such as uniformity, variance, entropy, mean intensity, product moment, correlation, inverse difference, contrast, information correlation, and modus) are crucial for identifying plant diseases and pests, especially at the micro (leaf) level, and the color cooccurrence method (CCM) is generally used for retrieving these features. Yao et al. [45] used a combination of shape and texture criteria to accurately categorize rice bacterial leaf blight, rice blast, and rice sheath blight. Some spatial metrics (i.e., landscape characteristics) obtained from RS images can detect the spatial distribution pattern of pests and plant diseases, and so provide a useful indication in their macrolevel monitoring.

Some efforts were made to use RS information to identify the habitat appropriateness of plant infections or pests in addition to the RS data that can be directly associated to damages produced by these pathogens or pests. Tasseled Cap Transformation (TCT)—based measures (i.e.,wetness, greenness, and brightness) related to plant vigor and soil moisture were found to be useful proxies of habitat suitability since stressed plants are more likely to be damaged by diseases and pests [46]. Important indicators for defining habitat status include TCT, VIs, and land surface temperature (LST). Several VIs (i.e., PRI, TVI, MSI, SAVI, MSAVI2, NDWI, and WI) are emphasized in the monitoring of mosquito larvae, tree woolly adelgids, spruce budworms, and wheat powdery mildew, all of which are related to plant biomass, pigment concentrations, and water content [47]. Some satellite products (e.g., MODIS-LST) and thermal bands of several satellites (e.g., Landsat TM, ASTER, HJ-IRS) provide LST, which can track the intensity of transpiration in plants [48].

Using satellite remote-sensing data, crop maps for polyphagous insects such as *Bemisia tabaci* and *B. argentifolii* can be generated [49]. These maps are very helpful for area-wide pest control. Pest migrating insects such as *Spodoptera exempta*, *Heliothis zea*, and *H. virescens* were tracked using specialized entomological radars. It may be possible to examine environmental conditions and predict desert locust activity using satellite remote sensing.

3.7 Spatially variable rate technology for spraying

The spatial variability in agricultural fields is managed through the variable-rate application of various inputs such as seed, fertilizer, insecticides, etc. The variable rate applicator consists of sensors, electrical controls, and mechanical drive systems that adjust the input rate in real time (VRA) etc. The variable rate applicator consists of sensors, electrical controls, and mechanical drive systems that adjust the input rate in real time (VRA).

Chemical application should be avoided in some field locations. The boom can be turned on and off in specific regions of the field using a boom control system, which can do this by analyzing a map of the area and activating or deactivating individual sections of the boom as needed. When the boom section enters a previously applied area, the controller can turn it off automatically to prevent overlaps [50]. Boom sections are automatically turned back on after leaving an area where they were applied, which also helps to prevent skips.

Remote sensing is a technique that will improve disease/weed scouting and result in better management decisions. It is possible for humans to visually discriminate between weed-free and weedy patches in a field and between distinct weed species based on plant features such as leaf shapes, diameters, inflorescence, etc. Some sprayers have a sensor that can tell the difference between soil and the green weeds growing between the rows of crops, and turn on or off the application device accordingly. WeedSeekers, a commercial product, uses a reflectance sensor to detect chlorophyll. Once the data is processed by the microprocessor and a predetermined threshold signal is exceeded (indicating the presence of weeds), the controller activates the spray nozzle. The unit allows for a significant reduction in the amount of chemical that needs to be applied in regions with varied levels of weed infestation (compared to uniform, continuous applications). Soil organic matter—based sensors can also be used for development of VRA of preherbicides because the amount of soil organic matter influences the effectiveness of some herbicides. Such a sensor can be used to automatically adjust herbicide rates without prescription maps or other inputs.

Weed-IT technology is accurate, fast, and easy to use for weed detection and its elimination. WEED-IT technology is on selective weed spraying and elimination for public paved areas from Wageningen University & Research (WUR), Netherlands [51]. By targeting weeds specifically and avoiding wasteful applications, spot spraying technology can reduce the need for pesticides in crop protection by as much as 70%. A variable-rate applicator developed and applied for patent by Punjab Agricultural University, Ludhiana (India) [52] can be used for the foliar spray of chemicals as shown in Figs. 3.3 and 3.4.

3.8 Opportunities and challenges

There could be many benefits of the application of geospatial technologies for the management of insect pests and diseases in crops, not only to the farmers but to

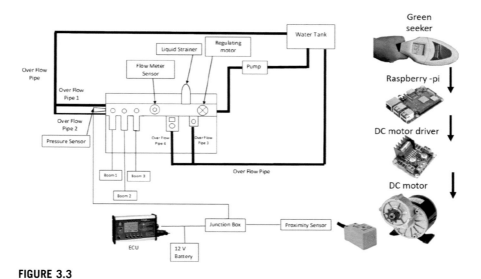

FIGURE 3.3

An illustration of developed real-time variable rate applicator for chemicals.

FIGURE 3.4

Operational view of developed variable rate application in rice and maize.

the environment also. The traditional method for collection of data for crop pests and diseases is challenging and expensive. Thus, comprehensive, real-time measures about the presence of pests and diseases, where, and to what severity; would be highly beneficial. Geospatial technologies provide an opportunity for a step-change in crop pest and disease surveillance around the world, particularly in agriculture-based economic countries [53].

The temporal lag that can occur between the capture of data and its subsequent distribution to the operator is one of the issues that satellite platforms and, to a lesser extent, drones and unmanned aircraft have faced. This lag can last for many days. When a satellite only passes over an area once every 5–6 days, or even longer, and clouds frequently cover the ground in that area, it can be challenging to do timely sensing and intervention. Unmanned aerial vehicles (UAVs) offer greater adaptability because they may be readily scheduled to maximize data collection.

In the future, unmanned aerial vehicles (UAVs) will be used more frequently in agriculture, since they are more cost-effective than satellite or manned aerial platforms and allow for a higher temporal density of remotely sensed data. Research is required across platforms to determine what factors lead to crop stress and how to directly quantify nutrient deficits in the absence of reference treatments.

Information gathered by remote sensing can be quite helpful in pinpointing problem areas in fields. It is difficult to determine the origin of such problems using spectral reflectance data; however, this is an active area of study. Current practices often involve using high-resolution remote sensing to locate problematic areas in fields, followed by on-the-ground scouting to determine the root cause. It is possible to employ need-based treatment to address problems after they have been mapped using remote sensing data. The convergence of distant sensing, vehicle-mounted, proximity sensing, and in-situ sensing is anticipated to continue. Control of inputs like fertilizer, irrigation, and pesticides will be more timely and automated due to the full integration of sensor information from these many platforms into management systems.

Recently Syngenta Crop Protection launched World's first commercial digital diagnosis and mapping solution to diagnose infestations of plant-parasitic nematodes in soybean crops through satellite images [54]. Brazilian soybean farmers will use the tool to fight pests causing up to 30% yield loss. Digital tool is the result of a multiyear collaboration of Syngenta with Swiss AgTech startup Gamaya SA.

Despite the hopeful advancements made in the past few decades in the monitoring of plant diseases and pests, there are still some issues that prevent the techniques from being used in real-world situations. Future trends will be influenced by research on finding answers to these problems. Monitoring plant diseases and insect pests involves four different aspects [51].

Currently, the majority of assessment studies or applications are carried out in experimental fields where there is already knowledge about the types of diseases/pests or other stresses [52,54,55]. The first challenge is to accurately identifying a specific disease or pest under actual field conditions, when many biotic and abiotic stresses are in crops may occur at the same time. Deep learning algorithms and other cutting-edge algorithms might be useful in this process. Additionally, the background information such as their geographic distribution, favorable habitats, soil types, and climate conditions as well as a network of pertinent supplementary data like soil data, meteorological data, and data from some wireless sensor networks may also be useful for accurate prediction of pests and diseases.

The second challenge is the early identification of plant disease or pest. Plant disease and pest monitoring using remote sensing is only possible when indications are completely manifested and that may be too late to apply the preventive measures. It is crucial to further leverage the viability of fluorescence, SAR, thermal, and LIDAR-RS observations and combine them with the highly developed VIS-NIR observations in order to increase the detectability of disease or pest at an early stage. Additionally, it would be worthwhile to try using multiangular RS to improve the capacity to identify the pest and disease at lower canopy levels [6].

The third challenge is to continuously monitor the changes in the spread/outbreak of the diseases or pests at high resolution. The remote sensing must have an adequate level of resolution across all spatial, spectral, and temporal dimensions to do this. In order to solve this issue, it is critical to study the viability of merging high-resolution satellite photos with unmanned aerial vehicle (UAV) photographs to produce successive time-series data. A substantial attention is also placed on the blending of optical and radar data.

The sharing of data and information is the fourth challenge. Taking into account the spread and outbreaks of plant diseases and pests (such a locust invasion) is a global process. Pest and disease identification and epidemic control are two fields that benefit greatly from international cooperation, both in the lab and in the field. For instance, farmers or extension agents in different regions/countries should be mobilized to use a smartphone app to log the existence and severity of pests and diseases in their fields. The pooled data is readily available to facilitate effective data mining and model training using advanced algorithms.

This may be taken as opportunity in order to enable and plan the experiments, data collection, modeling, and idea sharing at a continental or global scale; it is anticipated that the appropriate international projects and observation networks will be set up, just as it has already been done for creating soil spectral libraries at the global level.

References

[1] Mandal D, Ghosh PP, Dasgupta MK. Appropriate precision agriculture with site-specific cropping system management for marginal and small farmers. Plant Sci Rev 2013;121:1—6.

[2] Wojtowicz M, Wójtowicz A, Piekarczyk J. Application of remote sensing methods in agriculture. Commun Biometry Crop Sci 2016;11:31—50.

[3] Elbattay A, Hashim M. Induced em-spectral response for an early detection of vegetation (crop) biotic stress. Natural Science and Natural Heritage; 2010.

[4] Li H, Zhao C, Yang G, Feng H. Variations in crop variables within wheatcanopies and responses of canopy spectral characteristics and derived vegetationindices to different vertical leaf layers and spikes. Remote Sens Environ 2015;169:358—74.

[5] Jiannong X, Harmon CL, Vegot P, Huafeng J. Tracking pest and plant disease through space and time using geo-tagged digital images. In: Proceedings of conference on 7th world congress on computers in agriculture. Reno, Nevada. 22—24 June 2009; 2009.

[6] Gougherty AV, Nutter Jr FW. Impact of eradication programs on the temporal and spatial dynamics of Plum pox virus on Prunus spp. in Pennsylvania and Ontario, Canada. Plant Dis 2015;99:593—603.

[7] Ploetz R, Freeman S, Konkol J, Al-Abed A, Naser Z, Shalan K, et al. Tropical race 4 of Panama disease in the Middle East. Phytoparasitica 2015;43:283—93.

[8] Aguilar CH, Lasalita-Zapico F, Namocatcat J, Fortich A, Bojadores RM. Farmers' perceptions about banana insect pests and integrated pest management (IPM) systems in SocSarGen, Mindanao, Philippines. IPCBEE 2014;63:22—7.

[9] Idris NH, Said MN, Fauzi MF, Yusri NAM, Ishak MHI. A low-cost mobile geospatial solution to manage field survey data collection of plant pests and diseases presented in IEEE workshop on geoscience and remote sensing. Kuala Lumpur, Malaysia: IWGRS2015; 2015.
[10] Anonymous. Web based decision support system for late blight disease of potato crop. 2011. https://www.pau.edu/potato.
[11] Saleem RM, Kazmi R, Bajwa IS, Ashraf A, Ramzan S, Anwar W. IoT-based cotton whitefly prediction using deep learning. Scientific Programming; 2021. p. 8824601. https://doi.org/10.1155/2021/8824601.
[12] Stafford JV. Implementing precision agriculture in the 21st century. J Agric Eng Res 2000;76(3):267−75.
[13] Franke J, Menz G. Multi-temporal wheat disease detection by multi-spectral remote sensing. Precis Agric 2007;8:161−72.
[14] Gebbers R, Adamchuk VI. Precision agriculture and food security. Science 2010;327:828−31.
[15] Fiorani F, Schurr U. Future scenarios for plant phenotyping. Annu Rev Plant Biol 2013;64:267−91.
[16] Guo Q, Zhu Z. Phenotyping of plants. Encyclopedia of Analytical Chemistry; 2014. p. 1−15.
[17] Oerke EC, Mahlein AK, Steiner U. Proximal sensing of plant diseases. In: Gullino ML, Bonants PJM, editors. Detection and diagnostic of plant pathogens, plant pathology in the 21st century. Netherlands: Springer Science and Business Media; 2014; 2014. p. 55−68.
[18] Walter A, Liebisch F, Hund A. Plant phenotyping: from bean weighing to image analysis. Plant Methods 2015;11:14.
[19] Shannon DK, Clay DE, Kitchen NR. Pest measurement and management. Precis Agric Basics 2018:93−102.
[20] Dhaliwal GS, Arora R, Dhawan AK. Crop losses due to insect pests and determination of economic threshold levels. In: Singh A, Trivedi TP, Sardana HR, Sharma OP, Sabir N, editors. Recent advances in integrated pest management. New Delhi: National Centre for Integrated Pest Management; 2003.
[21] Anonymous. Clemson university pest identification site. 2010. (http://entweb.clemson.edu/pesticid/saftyed/pstident.htm).
[22] Anonymous. Online insect identification. West Virginia University Extension; 2010. https://extension.wvu.edu/lawn-gardening-pests/pests#id-guide.
[23] Anonymous. The BugGuide. 2010. http://bugguide.net.
[24] Rafoss T, Skahjem J, Johansen JA, Johannessen S, Nagothu US, Floistad IS, et al. Improving pest risk assessment and management through the aid of geospatial information technology standards. NeoBiota 2013;18:119−30.
[25] Aggarwal N, Sandhu GS, Suri KS. Economic threshold levels of major insect-pests for judicious use of insecticides. Prog Farming 2022;58(06):12.
[26] Singh S, Pandher S, Singh K. Mating disruption technology for pink bollworm management in cotton. Prog Farming 2022;58(06):13−4.
[27] Duarte F, Calvo MV, Borges A, Scatoni IB. Geostatistics applied to the study of the spatial distribution of insects and its use in integrated pest management. Rev Agron Noroeste Argent 2015;35(2):9−20.

[28] Zhang J, Huang Y, Pu R, Gonzalez-Moreno P, Yuan L, Wu K, et al. Monitoring plant diseases and pests through remote sensing technology: a review. Comput Electron Agric 2019;165:104943.

[29] Huang J, Arthanareeswaran G, Zhang K. Effect of silver loaded sodium zirconium phosphate (nanoAgZ) nanoparticles incorporation on PES membrane performance. Desalination 2012;285:100−7.

[30] Zhang J, Pu R, Huang W, Yuan L, Luo J, Wang J. Using in-situ hyper-spectral data for detecting and discriminating yellow rust disease from nutrient stresses. Field Crop Res 2012;134:165−74.

[31] Oumar Z, Mutanga O. Using WorldView-2 bands and indices to predict bronzebug (*Thaumastocoris peregrinus*) damage in plantation forests. Int J Rem Sens 2013;34:2236−49.

[32] Adelabu S, Mutanga O, Adam E. Evaluating the impact of red-edge band from Rapideye image for classifying insect defoliation levels. ISPRS J Photogramm Remote Sens 2014;95:34−41. https://doi.org/10.1016/j.isprsjprs.2014.05.013.

[33] Yuan L, Huang Y, Loraamm RW, Nie C, Wang J, Zhang J. Spectral analysisof winter wheat leaves for detection and differentiation of diseases and insects. Field Crop Res 2014a;156:199−207.

[34] Xu HR, Ying YB, Fu XP, Zhu SP. Near-infrared spectroscopy in detecting leaf minor damage on tomato leaf. Biosyst Eng 2007;96:447−54.

[35] Apan A, Held A, Phinn S, Markley J. Detecting sugarcane 'range rust' disease using EO-1 Hyperion hyperspectral imagery. Int J Rem Sens 2004;25:489−98.

[36] Anonymous. Package of practices for kharif crops. Ludhiana (India): Punjab Agricultural University; 2022.

[37] Naidu RA, Perry EM, Pierce FJ, Mekuria T. The potential of spectralreflectance technique for the detection of Grapevine leafroll-associated virus-3 in twored-berried wine grape cultivars. Comput Electron Agric 2009;66:38−45.

[38] Luo J, Huang W, Zhao J, Zhang J, Zhao C, Ma R. Detecting aphid density ofwinter wheat leaf using hyperspectral measurements. IEEE J Sel Top Appl EarthObs Remote Sens 2013;6:690−8.

[39] Prabhakar M, Prasad YG, Thirupathi M, Sreedevi G, Dharajothi B, Venkateswarlu B. Use of ground based hyperspectral remote sensing for detection of stress in cotton caused by leafhopper (Hemiptera: cicadellidae). Comput Electron Agric 2011;79(2):189−98.

[40] Bürling K, Hunsche M, Noga G. Presymptomatic detection of powdery mildewinfection in winter wheat cultivars by laser-induced fluorescence. Appl Spectrosc 2012;66:1411−9.

[41] Kuckenberg J, Tartachnyk I, Noga G. Detection and differentiation of nitrogen deficiency, powdery mildew and leaf rust at wheat leaf and canopy level by laser-induced chlorophyll fluorescence. Biosyst Eng 2009;103:121−8.

[42] Iqbal MJ, Goodwin PH, Leonardos ED, Grodzinski B. Spatial and temporal changes in chlorophyll fluorescence images of *Nicotiana benthamiana* leaves following inoculation with *Pseudomonas syringae* pv. tabaci. Plant Pathol 2012;61:1052−62.

[43] Stoll M, Schultz HR, Baecker G, Berkelmann-Loehnertz B. Early pathogen detection under different water status and the assessment of spray application invineyards through the use of thermal imagery. Precis Agric 2008;9:407−17.

[44] Falkenberg NR, Piccinni G, Cothren JT, Leskovar DI, Rush CM. Remote sensing of biotic and abiotic stress for irrigation management of cotton. Agric Water Manag 2007; 87:23–31.
[45] Calderón R, Navas-Cortés JA, Lucena C, Zarco-Tejada PJ. High-resolution airborne hyperspectral and thermal imagery for early detection of *Verticillium* wilt of olive using fluorescence, temperature and narrow-band spectral indices. Remote Sens Environ 2013;139:231–45.
[46] Stratoulias D, Balzter H, Zlinszky A, Tóth VR. Assessment of ecophysiology of lake shore reed vegetation based on chlorophyll fluorescence, field spectroscopy and hyperspectral airborne imagery. Remote Sens Environ 2015;157:72–84.
[47] Yao Q, Guan Z, Zhou Y, Tang J, Hu Y, Yang B. Application of support vector machine for detecting rice diseases using shape and color texture features. Int Conf Eng Comput 2009:79–83.
[48] Zhang J, Pu R, Yuan L, Huang W, Nie C, Yang G. Integrating remotely sensed and meteorological observations to forecast wheat powdery mildew at a regional scale. IEEE J Sel Top Appl Earth Obs 2013;7:4328–39.
[49] Williams JP, Hanavan RP, Rock BN, Minocha SC, Linder E. Low-level *Adelges tsugae*, infestation detection in New England through partition modeling of Landsat data. Remote Sens Environ 2017;190:13–25.
[50] Prabhakar M, Thirupathi M, Mani M. Principles and application of remote sensing in crop pest management. In: Mani M, editor. Trends in horticultural entomology. Singapore: Springer; 2022. https://doi.org/10.1007/978-981-19-0343-4_5.
[51] Grisso RB, Alley M, Thomason W, Holshouse, Roberson GT. Precision farming tools: variable-rate application. PUBLICATION 442-505, produced by communications and marketing, college of agriculture and life sciences. USA: Virginia Polytechnic Institute and State University; 2011.
[52] Humburg D. *Variable rate equipment—TechnologyforWeedControl*. International plant nutrition Institute,Site-specific management guidelines, SSMG-7. Norcross, Ga. 1990. www.ppi-ppic.org/ppiweb/ppibase.nsf/webindex/articleA172CE4C8525696100631668C0F666E3orwww.agri-culture.purdue.edu/ssmc/.
[53] Anon. Website: https://www.weed-it.com/. Searched on 24.10.2022.
[54] Mahlein AK. Plant disease detection by imaging sensors: parallel and specific demands for precision agriculture and plant phenol-typing. Plant Dis 2016:241. https://doi.org/10.1094/PDIS-03-15-0340-FE [2016 The American Phyto-pathological Society].
[55] Ng M. Syngenta-launches-worlds-first-commercial-digital-tool-detect. 2022. https://www.syngenta.com/en/company/media/syngenta-news/year/2022/syngenta-launches-worlds-first-commercial-digital-tool-detect.

CHAPTER 4

Application of unmanned aerial vehicles in precision agriculture

Muhammad Naveed Tahir[1], Yubin Lan[2], Yali Zhang[2], Huang Wenjiang[3], Yingkuan Wang[4], Syed Muhammad Zaigham Abbas Naqvi[1,5]

[1]Department of Agronomy, PMAS-Arid Agriculture University Rawalpindi, Rawalpindi, Punjab, Pakistan; [2]National Center for International Collaboration Research on Precision Agriculture Aviation Pesticides Spraying Technology, South China Agricultural University, Guangzhou, Guangdong, China; [3]Aerospace Information Research Institute, Chinese Academy of Sciences, Beijing, China; [4]Chinese Academy of Agricultural Engineering Planning and Design, Beijing, China; [5]College of Mechanical and Electrical Engineering, Henan Agricultural University, Zhengzhou, Henan, China

4.1 Introduction

Early crop yield predictions play a critical role in the development of economic policies not only at the national level but also at the regional/global scale, in order to design the agricultural strategies and forecast a nation's GDP [1,2]. Conventional techniques for anticipating and estimating crops yields consisted of destructive sampling, which was not only labor-intensive but also time consuming. These yield estimates based on crop area are not entirely accurate and reliable. The United Nations is working to achieve food security and zero hunger by 2030, and if this goal is not met, the number of malnourished people could be around 690 million [3,4]. These scenarios aid in mitigating and assessing the threats to food security in timely manner; the most significant of these threats is the inability to accurately predict crop yields prior to the harvest. Developing nations are already in the process of enhancing food security and crop yield projections for increased crop production [5,6].

Precision agriculture is the remarkable technique that not only increases crop production efficiency by minimizing resource inputs and optimizing nutrient and water use, but also increases crop yield and protects the environment [7]. The crop monitoring and the assessment of crop can easily be performed using remote sensing (RS) techniques. The RS collects the air and space-borne data to monitor crops based on the optical and biophysical properties of plants [8]. This agricultural remote sensing facilitates nondestructive, real-time diagnosis of crop growth and yield predictions prior to crop harvesting, which is a time and cost-efficient alternative to the conventional methods. Satellite based remote sensing can also derive the important indices for crop growth monitoring such as leaf area index (LAI) and leaf dry matter (LDM), but spatial and temporal resolution limitations of satellite images can limit the precision and accuracy of growth monitoring and yield estimations [9].

4.2 Types of UAVs

UAVs are classified into several types based on their intended use, such as multirotor, single rotor, fixed wing, fixed-wing multirotor hybrids, and helicopters. Their characteristics vary depending on their use, such as long and short flight times, the ability to gain different altitudes with good hovering abilities.

4.2.1 Multirotor UAVs

Multirotor UAVs are ideal for detecting aphids or pests in many crops. Multirotor UAVs have more than two rotors, and spinning blades are attached to the rotor to create efficient lift (Fig. 4.1). Changing the speeds of the rotors can also allow the drone to move horizontally.

4.2.2 Fixed-wing UAVs

Fixed-wing UAVs are most likely to the normal aircraft (Fig. 4.2). These drones, unlike multirotor, do not require as much energy to stay in the air because they cannot hover in one place. This ability allows them to distinguish the fly for longer periods of time and capture images across the large area of the field.

4.2.3 Single-rotor UAVs

These UAVs resemble helicopters in both design and structure. Although they are referred as single rotors, the design includes two, one at the top and other at the tail (Fig. 4.3). These rotors have a longer flight time and can hover in place for aerial imaging and precision spraying. However, they are usually powered by the gas engines. These rotors are much more steerable and efficient than multirotor. Large sized blades of single-rotor UAVs also pose operational risks and necessitate special training.

FIGURE 4.1

Quadcoptor UAV.

FIGURE 4.2

Fixed-wing UAV.

FIGURE 4.3

Single-rotor UAV.

4.2.4 Fixed-wing multirotor hybrid UAVs

The hybridization of fixed wing and multirotor UAVs laid the foundation of these types of UAVs which carry the ability to not only perform the vertical takeoff and land but can perform hovering for spot and aerial imaging. These fixed-wing multirotor hybrid UAVs have ability of long flight time and can stay in the air much longer, comparatively. The UAV in Fig. 4.4 is an example of such hybrid drones that possesses multipurpose characteristics of imagery data acquiring, precision spraying, and surveillance abilities.

4.2.5 A hybrid fixed-wing multirotor UAV

The four types of UAVs mentioned above are commonly used by scientists and for commercial purposes, but there is also a unique and novel flexible membrane wing UAV type that can perform the targeted flight in severe and windy conditions (Fig. 4.5). As the name suggests, these UAVs are made up of flexible membrane wing, which has the added benefits of carrying a larger load, easy storage capacity, better control, and maneuverability.

FIGURE 4.4

Fixed-wing multirotor hybrid UAV.

FIGURE 4.5

Multirotors with hexa wing UAV.

4.3 Recent advances in the UAV technology

Microtechnology advancements have resulted in the widespread use of UAV technology in many fields of science, as they can compete with high-altitude systems such as satellites for data collection and can even collect data from small plots of land [10]. The remotely sensed data collected has high spatial and temporal resolutions, which aid in the generation of nondestructive, precise, accurate, and pertinent estimations for crop biophysical attributes. Moreover, the data collected is free of observational errors that occur during conventional data collection, reducing errors, and emphasizing the importance of precision agriculture.

4.4 Literature review

4.4.1 Use of UAVs for crop health monitoring

The significance of UAVs touched its peak by the end of 20th century due to the use in agricultural remote sensing [11,12]. They are able to collect field data for different crop monitoring parameters including land cover and land use, growth monitoring, and yield estimations. The various types of UAVs like fixed wings and multirotor express various properties for data acquiring, for example, multirotor has an advantage of altitude adjustment during flight for convenient flight with improved accuracy and efficiency [13]. These characteristics of UAVs made them more preferable when compared to the ground remote sensing and high-altitude remote sensing. In addition, the more advanced and improved sensors of high accuracy and precision can be integrated with UAVs for remote sensing activities during field activity as well [14,15].

UAV-based remote sensing is critically significant in precision agriculture due to its exclusive edge of stability and high-resolution properties. UAVs make it possible to collect the accurate information for all the fields with high temporal and spatial resolutions [16]. The data obtained is in real time and nondestructive with efficient analytical abilities and possess the ability to predict the crop growth and yield from the vegetation imageries acquired during remote sensing activities.

Imageries taken from remote sensing methods using multispectral cameras are used to derive the broadband vegetation index, which are significant to derive the estimations for crop yields [17]. A good correlation among vegetation indices and crop yield can help to manage the strategical protocols for good productions. The spectral bands used to derive the vegetation indices are comprised of visible, near, and thermal infrared bands of remote sensing, which are fundamentally the reflectance obtained from the phenology and biophysical variables of the crop canopy [18]. These indices are the mathematical equations developed in a special arrangement of spectral bands to elaborate the vegetative characteristics of plants like vigor, health, canopy, chlorophyll contents, biomass, and radiation absorbed by the leaves. NDVI is the most used vegetation index to monitor the crop plants like crop simulations, and yield estimations, and environmental conditions like dry land and land degradations [19].

The monitoring of crop plants and environmental conditions has been widened by the advent of modern computational advancements and increased UAV operations in the modern agricultural techniques. UAV operations have now extensively been applied to gather the aerial imageries and field data since last few years due to their characteristics features of flexibility and weather independence when compared to the conventional airborne aerial surveys [20]. Micro-UAVs with a total weight less than 5 kg are considered to be more reliable, flexible, affordable source of agricultural surveys with accurate geoinformation and to be able to complete the photogrammetric assignments. With the advent of high-resolution imageries derived from the complex high-end multisensory system, the models were reconstructed to derive the spectrum ranging from various wavelengths [21]. The integration of cost-effective and lightweight RGB cameras with mini and micro-UAVs has proved its efficiency for fast and valuable agricultural monitoring and detection system at low altitude [22]. Additionally, these UAVs are the best alternative source of traditional and high-altitude sensing technologies used for farmland data gathering.

The countries practicing the complex farming structures like multiverities and small-scale plantations face the difficulties of farmland data gathering even with the conventional techniques. The UAVs like multirotor having the ability of vertical takeoff and landing can gather the accurate crop information and monitor the field activities in a small field environment [23]. With the extra benefit of hovering ability, the UAV can obtain the clear and accurate imageries to derive the field crop information. Assistance in wheat breeding and multiple other parameters like crop lodging area, leaf area index, and canopy temperature were experienced by using the multirotor UAV mounted digital camera and multispectral thermal imaging camera [24]. The hyperspectral data for inversion estimation of cotton was also obtained by using the spectral sensor integrated with eight-rotor UAV [25]. The same UAV can also be used to estimate the yield of soybean by acquiring the imagery at critical growth stages using the hyperspectral sensor [26].

4.4.2 UAVs for yield estimation

Yield estimation studies and crop monitoring has continuously been performed since long time using UAV technologies. The other sensory platforms and other remote sensing methods carry certain limitations like the lacking abilities of preprocessing and noise removal from the imageries. These lacking abilities are overcome by the development of improved and accurate predictive models like leaf area index and leaf dry matter index, which can thoroughly express the physiological parameters and crop growth status precisely [27]. Literature has already expressed the significance of these indices to monitor the growth indicators using UAV methods such as digital camera mounted on quadrotor UAV analyzed and developed the significant correlation of LAI and canopy coverage with three models [28]. Another significant relation of LAI and wheat growth stages was developed by acquiring the spectral data using imaging spectral sensor integrated with multirotor UAV [29]. The feasibility of LAI to monitor the different verities of crop was also carried out by using UAV mounted with hyperspectral camera [30].

The improvements and expansion of UAV technologies have provided a significant new pathway to the field of remote sensing. Nondestructive agricultural monitoring and rapid yield estimations have become more precise and accurate in wheat by using the multispectral camera on UAV platform [31]. Imageries obtained at different crop growth stages like jointing, heading, and filling stages, were used to measure yield by deriving the nine linear models for vegetation indices [32]. The same analysis for yield estimation of rapeseed was performed by deriving the NDVI index from the imageries obtained at early flowering period using multirotor UAV and multispectral camera [33]. Currently, the research process of using multisensor UAV platform to obtain crop canopy spectral image data for crop yield estimation has been preliminarily determined. Therefore, more research focuses on the selection of vegetation index and the improvement of yield accuracy estimation model.

The grain yield of crop and quality of the grain specifically for protein contents that can vary from crop to crop can be optimized by appropriate field managements and adopting suitable agronomic practices at the critical growth stages of crop plants [34]. The understanding of these factors is critically important for the farmers to monitor at every stage of the crop in real time during its development in the field and take managerial on-spot decisions. The reason that makes the remote sensing an important tool for agricultural purposes is that the reproductive growth of crops after flowering is closely related to grain yield. In recent years, many studies on crop growth have shown that the cumulative NDVI value of multitemporal satellite remote sensing images after flowering has a good effect in relation to the crop yield [35].

The UAV remote sensing is an alternative efficient technical tool to justify the assistance and managerial decision-making support for field activities. In an experiment of wheat crop, from the multitemporal UAV othomosaic imageries, the visualization of the rapid changes in the growth conditions and discern the canopy greenness of wheat fields through image interpretation was obtained. Field changes in wheat tiller density can also be seen, and the manifestation of lodging can be preempted to prescribe the variable fertilizer applications to avoid or mitigate the rate of lodging by decreasing the amount of fertilization around areas with excessive tiller density. Acquisition of othomosaic imageries before the harvest of the crop can enable to take the predictive measures and guide the drivers of combine harvesters or automated harvesting vehicles to adjust their operating speed based on specific lodging frequency in the field [36].

Quantification of the height and the vigor, known as the canopy traits, was performed using the UAV and MGP imaging for wheat trails. The results for both traits were found to be significant with the UAV and MGP imaging; MGP imaging was found to provide better predictions of height, while UAV imaging was able to quantify the better estimates of canopy vigor. Although both imaging methods were found to be providing accurate quantifications, yet the best suited system for field phenotyping depends on the applications in the fields. UAV enables to get adequately accurate estimations of canopy-based traits and cover the larger field pieces of land rapidly. Contrarily, the MGP imaging is not labor-efficient and does not provide rapid outputs; but the data collection from the field is comprised of the detailed canopy structure to provide the traits analysis lowest to the plant level [37].

UAV imaging for RGB can also provide the estimation of plant density in the field crop. The field conditions vary under normal situation due to the environmental fluctuations. The same experiment conducted for wheat trails to retrieve the data for plants density resulted in the range of 79–388 plants per meter square and the root mean square error 21.66–52.35. The uncertainties were also considered to get the target accuracy which was about 10%. To get the highest accuracy of plant density, the spatial resolutions need to be reached at least 0.40 mm which can easily quantify the number of plants per image in the field of overlapped plantings. The required low altitude flight can lead to the potential source of problems for photogrammetry techniques used to retrieve the precise position of the camera when acquiring images. On the other hand, using an onboard centimeter-accurate geolocation system integrated with inertial measurement unit camera orientation information, the image can be projected directly onto the ground with sufficient accuracy: due to the low altitude, the camera orientation from the inertial measurement unit will translate into limited positioning error. Due to the relatively simple and automated methods for collecting and processing images, plant density estimation can be applied to large samples [38].

4.4.3 Role of UAVs in pest detection

The outbreaks of the arthropod pests are not easily predictable and also it does not spread in the field in uniform pattern. To detect early and treat the pest outbreaks comes under the management of pests, which requires to take necessary measures to avoid crop failure and yield losses. Pest scouting and monitoring needs a lot of time and labor during the sampling process manually within the field. Therefore, we believe that a significant research task related to maximize the sustainability of modern agricultural pest management is the development and promotion of improved crop monitoring programs.

These crop monitoring programs are now using the nondestructive methods like imaging techniques that are quite effective in crop monitoring and precision applications. These methods use different sensors for sensing and quantifications and these sensors can be easily integrated on the instruments moving within the fields or can be combined with drones or unmanned air robots (Fig. 4.6). Canopy reflectance values attained and processed by sensing drones may be communicated to generate the digital map to direct the second driving drone, to provide solutions for the detected pest hotspots.

UAVs have the benefit of high accuracy over manned crafts due to the flight ability at lower heights and collection of high spatial resolution imageries that can avoid the total numbers of mixed pixels. Moreover, due to the frequent flight abilities that have high-cost benefits over crafts and satellites because they don't require launching pads and high fuel expenses. In addition, due to rapid flight ability they are not dependent on revisiting times like satellites and can monitor the field whenever required [39,40].

The use of UAV methods for crop monitoring is determined as nondestructive technique and can determine the plant stresses early before changes in the field can be seen by the naked eye. The remote sensing methods can usually read the spectrum that falls under the range of visible light (400−700) nm and near infrared spectrum (700−1400) nm. Most of the studies have referred the near infrared spectrum to study the plant stress or vegetational changes in the field [41]. Specific stress factors like in the case of arthropod attack, the photosynthetic efficiency of plants become critically low thus causing variations in reflectance values of spectral wavelengths of leaves. Drones can be mounted with different sensors, like RGB sensors;

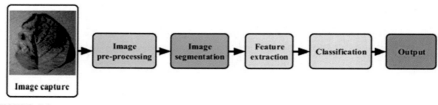

FIGURE 4.6

Schematic diagram for disease detection using RGB camera.

hyperspectral, thermal, and multispectral cameras for different purposes of aerial sensing. The wavelengths of spectral bands can vary for different sensors ranging from 3 to 12 broad bands and up to 100 narrow bands.

4.4.4 Role of UAVs for disease detection

Drones when introduced in commercial facilities were only used for spraying chemicals on the crops, but their developments have brought them to use for aerial remote sensing platform by integrating with cameras and sensors. These sensors can take imagery with varying spectral bands ranging from visible light photographs to multispectral imageries. The analysis of these imageries can be used to monitor the health of plants and predict the various weeds, diseases, and other stresses (Fig. 4.7).

4.5 The UAV-based precision agriculture chemical spraying application

The need to secure the food supply for rapidly increasing population is also a challenge that has to be addressed using advance technologies like UAVs. Even though agriculture is a major sector of economy, it still has not developed enough to adapt latest technologies. The critical factors which severely affect the quality and yield of the crops are diseases and weeds. Their effective remedy determined by researchers is chemical spraying. A major reducing factor in yield of agriculture is pests. It is estimated that 20%–40% of global crop production are lost to the pests [3]. UAV drones can be used efficiently to control pests. They were initially used for chemical spraying [42]. Traditional methods of pesticides spraying for the control of insect/pest not only are time consuming, but they are also hazardous to the human and as well to the environment. The operator of the knapsack sprayer is always at the risk of pesticide poisoning. The plant-protection flight procedure has also been reformed from conventional artificial spraying to mechanical spraying [43].

FIGURE 4.7

UAV equipped with sensor for crop health monitoring.

With the advancement in the Unmanned Aerial Vehicles (UAVs) technologies, there is need to evaluate and adopt advance technology for pest control. With the advancement in technologies there is a need to adopt advance technologies for pest control in various crops. There has been a very rapid increase in use of UAVs in agriculture over the past few years. UAVs can be used for crop management and monitoring; they can provide accurate imaging; they can be used for crop dusting and spraying as well. UAVs can spray large areas within no time and they can be used as a potential solution to deal with the decreasing labor availability in rural areas. Moreover, UAVs hold the potential to mitigate the health hazards that are caused during the manual spraying of the chemicals. Therefore, using UAV drone sprayer is a better option as compared to traditional spraying practices. The main part of the UAV sprayer is frame which is needed to be strong enough to carry payload but light weight as well. Other components of UAV sprayer include rotors to provide thrust for the propelling of the UAV, IMU sensors to measure the attitudes of the UAV, microcontroller an on-board processor to collect data, implement control algorithm, drive actuators, and communicate with ground stations. It also contains a battery as a power source for whole UAV, radio control transmitter, GPS, sensors, and camera vision subsystem. The spray system comprised of four fundamental mechanisms: a boom arm with mounted spray nozzles, a tank to store the spray material, a liquid gear pump, and a mechanism to control spray activation.

Farmers in developed countries have started using UAV drones for crop treatments and to improve the farm outputs. UAVs for agriculture have expanded rapidly in recent years because of their incredible potential practices in precision agriculture. Similar to conventional aerial mapping technique that involves using manned aircraft, UAV system uses unmanned aerial vehicle that carries a camera (imaging sensor) and serves as an airborne platform for taking images [44]. The traditional manned aircraft imagining system is expensive and is not feasible in most of the scenarios whereas the UAV system works in almost all the cases and is relatively cheap. As compared to satellite technology, using the drones in smart farming and precision agriculture is a better option as it can provide a better bird eye view of the field to the farmers while staying closer to the surface [45]. Moreover, UAV drones save up a lot of time since they can provide an overall survey of the field quickly and spot abnormalities, pests, and diseases. Navigation provided by generating the waypoints at ground station, drones become automatic navigators and move to the waypoints to hover and acquire the live field image data (Fig. 4.8).

UAV when combined with the sprayer system can transform into the platform for pesticide spraying for their management and control. This platform is comprised of automated drone and sprinkling system attached with multispectral camera. The composition of sprinkling system has a pesticide tank with nozzles to sprinkle the pesticides in downwards direction. Before the spraying process, the attached camera scans and takes imageries of the field to analyze the field conditions using NDVI analysis map. The analysis further evaluates and directs the needs of pesticide for field applications. This UAV-based spraying system (Fig. 4.9) is used for accurate site-specific application of pesticides on a large crop field [46].

FIGURE 4.8
UAV-based agrochemical spraying.

FIGURE 4.9
Precise UAV-based precision spraying at farmer's field.

4.6 Artificial intelligence for UAVs

The Artificial Intelligence (AI) field with its robust learning abilities has become a primary technique for tackling the challenges related to agriculture. Recent advances in AI techniques (Deep Learning, Machine Learning, and Machine Vision) are frequently being used in agriculture for crop management and decision-making. Smart technology for efficient farming-related implementations is UAV (Unmanned Aerial Vehicles). Aerial surveillance of UAV farms in agriculture allows for crucial decision-making on monitoring of crop. The accuracy and reliability of aerial imagery—based research has been further improved by developments in deep learning models. Remote crop assessment applications such as plant categorization and aggregation crop counting, yield estimate and comparison, weed identification, disease detection, crop mapping, and nutrient deficiencies, among others, may accommodate various types of sensors (spectral cameras, RGB) on UAVs [47]. The recent studies explore adapting deep learning to UAV imagery for smart farming practices. Depending on the requirement, these studies were categorized into six main classes, which include: identification of vegetation, classification and segmentation, crop estimation and yield expectations, crop mapping, crop disease and detection of weeds, and detection of inadequate intake. An in-depth comprehensive analysis of each research provided detailed artificial intelligent approaches in the vast agricultural subdomain, so the readers are able to capture the multidimensional advancement of agro-intelligent systems over the past 34 years [48].

Kerkech et al. [49] proposed novel technique for identifying vine disease using multimodal UAV pictures that has been suggested in this work, which is based on improved image detection and deep learning algorithms (visible and infrared ranges). There are three phases to the process. The first one is to take pictures that are obtained based on interest points detector. Research showed that system to be working very efficiently for vine detection using infrared spectra. It described that early detection of vine disease and mapping can be possible with reasonable accuracy.

Su et al. [50] investigated that this research is to combine a reduced five-band multispectral camera (RedEdge), a flying low platform, and reducing machine learning algorithms to autonomously detect wheat crop stress by yellow rust. Whenever yellow rust disease appeared, testing results demonstrate that the proposed system showed promising results of deep learning model for rust detection system.

4.7 Challenges and limitations of UAV technology adoption

UAVs with their significant advantages are currently being used even in high-risk conditions to acquire the images. But where there are some benefits, there are also some limitations attached with the adoption of UAVs. If the cost-efficient UAVs are used in the field, they face difficulties to integrate with the sensors due to their variable size, dimensions, and weight. Due to their less power engines, the UAVs face difficulties to reach beyond a specific height. Moreover, there are some situations faced by operators, e.g., path planning, high speed with low resolution, real-time data downloading functions, size, and payload of UAV for evading the bottle neck.

A real and critical problem is the adoption of drone technology by following the country laws. Most of the countries don't allow the use or application of drones due to the security issues. Although some regions of the world have adopted UAVs in agriculture, they are in limited numbers. Most of the African countries and some of the Asian countries including Pakistan need to revise their policies to opt for UAV operations in agriculture field.

4.8 Conclusion

Agriculture is the only part that can keep the ecosystems alive on the Earth. Every living organism is dependent on agriculture to thrive. This chapter explores the different aspects of unmanned aerial vehicles application in agriculture particularly focused on crop monitoring, growth estimation, yield predictions, early detection of diseases and insects/pests attack using the multispectral and thermal camera integrated on multirotor UAV. The UAV can overcome the losses of yield and reduce the cost and provide better field managements and help to mitigate the food security disasters. Not only this, but UAVs possess the vital characteristics of flying at lower heights and with flexible environmental conditions posing a new trend in remote sensing technology. They can generate the data with high resolutions, frequencies, and multisource ability. Further, this chapter elaborates the nondestructive and relevant applications of different types of drones in agricultural activities. The integrated technologies like field analysis and automated decision-making system can further solve the specific problems of the farmers and boost the precision applications of UAVs during the crop growth season. The recent advances in Artificial Intelligence (Machine Learning/Deep Learning) further widen the scope of UAVs for better detection and precision management—based application in future agriculture.

References

[1] Goel RK, Yadav CS, Vishnoi S, Rastogi R. Smart agriculture—urgent need of the day in developing countries. Sust Computing: Informat Syst 2021;30:100512.

[2] Talasila V, Prasad C, Reddy GTS, Aparna A. Analysis and prediction of crop production in Andhra region using deep convolutional regression network. Int J Intell Eng Syst 2020;13(5):1—9.

[3] FAO. The world is at a critical juncture: fao.org. Available from: https://www.fao.org/state-of-food-security-nutrition/2021/en/; 2021.

[4] United Nations. Goal 2: zero hunger: Un.org. 2021. Available from: https://www.un.org/sustainabledevelopment/hunger/.

[5] Lenaerts B, Collard BCY, Demont M. Review: improving global food security through accelerated plant breeding. Plant Sci 2019;287:110207.

[6] Guo Y, Chen Y, Searchinger TD, Zhou M, Pan D, Yang J, et al. Air quality, nitrogen use efficiency and food security in China are improved by cost-effective agricultural nitrogen management. Nature Food 2020;1(10):648—58.

[7] McBratney A, Whelan B, Ancev T, Bouma J. Future directions of precision agriculture. Precis Agric 2005;6(1):7—23.

[8] Stroppiana D, Migliazzi M, Chiarabini V, Crema A, Musanti M, Franchino C, et al., editors. Rice yield estimation using multispectral data from UAV: a preliminary experiment in northern Italy. IEEE international geoscience and remote sensing symposium (IGARSS); 2015 26—31 July 2015.

[9] Twumasi DA. Monitoring the impact of urea deep placement on rice (*Oryza sativa* L.) growth and grain yield using unmanned aerial systems technology (Drones). University for Development Studies; 2020.

[10] Huang Y, Thomson SJ, Hoffmann WC, Lan Y, Fritz BK. Development and prospect of unmanned aerial vehicle technologies for agricultural production management. Int J Agric Biol Eng 2013;6(3):1−10.

[11] Li D, Li M. Research advance and application prospect of UnmannedAerial vehicle remote sensing system. J Geomatics Inf Sci Wuhan Univ 2014;39(5):505−13.

[12] Naqvi SMZA, Tahir MN, Shah GA, Sattar RS, Awais M. Remote estimation of wheat yield based on vegetation indices derived from time series data of landsat 8 imagery. Appl Ecol Environ Res 2018;17(2):3909−25.

[13] Tahir MN, Naqvi SZA, Lan Y, Zhang Y, Wang Y, Afzal M, et al. Real time estimation of chlorophyll content based on vegetation indices derived from multispectral UAV in the kinnow orchard. Int J Precision Agric Aviation 2018;1(1):24−31.

[14] Li B, Liu R, Liu S, Liu Q, Liu F, Zhou G. Monitoring vegetation coverage variation of winter wheat by low-altitude UAV remote sensing system. Trans Chin Soc Agric Eng 2012;28(13):160−5.

[15] Tahir MN, Lan Y, Zhang Y, Wang Y, Nawaz F, Shah MAA, et al. Real time estimation of leaf area index and groundnut yield using multispectral UAV. Int J Precision Agric Aviation 2020;3(1):1−6.

[16] Naqvi SMZA, Awais M, Khan FS, Afzal U, Naz N, Khan MI. Unmanned air vehicle based high resolution imagery for chlorophyll estimation using spectrally modified vegetation indices in vertical hierarchy of citrus grove. Remote Sens Appl: Soc Environ 2021;23:100596.

[17] Narmilan A, Gonzalez F, Salgadoe AS, Kumarasiri UW, Weerasinghe HA, Kulasekara BR. Predicting canopy chlorophyll content in sugarcane crops using machine learning algorithms and spectral vegetation indices derived from UAV multispectral imagery. Rem Sens 2022;14(5):1140.

[18] Shen Y, Mercatoris B, Cao Z, Kwan P, Guo L, Yao H, et al. Improving wheat yield prediction accuracy using LSTM-RF framework based on UAV thermal infrared and multispectral imagery. Agriculture [Internet] 2022;12(6):892.

[19] Niu Y, Zhang L, Zhang H, Han W, Peng X. Estimating above-ground biomass of maize using features derived from UAV-based RGB imagery. Rem Sens 2019;11(11):1261.

[20] Dileep MR, Navaneeth AV, Ullagaddi S, Danti A, editors. A study and analysis on various types of agricultural drones and its applications. 2020 fifth international conference on research in computational intelligence and communication networks (ICRCICN); 2020 26−27 November 2020.

[21] Toth C, Jóźków G. Remote sensing platforms and sensors: a survey. ISPRS J Photogram Remote Sens 2016;115:22−36.

[22] Zhou M, Zhou Z, Liu L, Huang J, Lyu Z. Review of vertical take-off and landing fixed-wing UAV and its application prospect in precision agriculture. Int J Precision Agric Aviation 2020;3(4):8−17.

[23] Yang G, Liu J, Zhao C, Li Z, Huang Y, Yu H, et al. Unmanned aerial vehicle remote sensing for field-based crop phenotyping: current status and perspectives. Front Plant Sci 2017;8:1−26.

[24] Yang G, Li C, Yu H, Xu B, Feng H, Gao L, et al. UAV based multi-load remote sensing technologies for wheat breeding information acquirement. Trans Chin Soc Agric Eng 2015;31(21):184−90.

[25] Tian M, Ban S, Chang Q, You M, Luo D, Wang L, et al. Use of hyperspectral images from UAV-based imaging spectroradiometer to estimate cotton leaf area index. Trans Chin Soc Agric Eng 2016;32(21):102−8.

[26] Zhao X, Yang G, Liu J, Zhang X, Xu B, Wang Y, et al. Estimation of soybean breeding yield based on optimization of spatial scale of UAV hyperspectral image. Trans Chin Soc Agric Eng 2017;33(1):110–6.

[27] Wang S, Celebi ME, Zhang Y-D, Yu X, Lu S, Yao X, et al. Advances in data preprocessing for biomedical data fusion: an overview of the methods, challenges, and prospects. Inf Fusion 2021;76:376–421.

[28] Córcoles JI, Ortega JF, Hernández D, Moreno MA. Estimation of leaf area index in onion (Allium cepa L.) using an unmanned aerial vehicle. Biosyst Eng 2013;115(1): 31–42.

[29] Gao L, Yang G, Yu H, Xu B, Zhao X, Dong J, et al. Retrieving winter wheat leaf area index based on unmanned aerial vehicle hyperspectral remote sensing. Trans Chin Soc Agric Eng 2016;32(22):113–20.

[30] Aasen H, Burkart A, Bolten A, Bareth G. Generating 3D hyperspectral information with lightweight UAV snapshot cameras for vegetation monitoring: from camera calibration to quality assurance. ISPRS J Photogrammet Remote Sens 2015;108:245–59.

[31] Chen Z, Ren J, Tang H, Shi Y, Liu J. Progress and perspectives on agricultural remote sensing research and applications in China. J Remote Sens 2016;20(5):748–67.

[32] Zhu W, Li S, Zhang X, Li Y, Sun Z. Estimation of winter wheat yield using optimal vegetation indices from unmanned aerial vehicle remote sensing. Trans Chin Soc Agric Eng 2018;34(11):78–86.

[33] Gong Y, Duan B, Fang S, Zhu R, Wu X, Ma Y, et al. Remote estimation of rapeseed yield with unmanned aerial vehicle (UAV) imaging and spectral mixture analysis. Plant Met 2018;14(1):70.

[34] Qiong W, Cheng W, Jingjing F, Jianwei J. Field monitoring of wheat seedling stage with hyperspectral imaging. Int J Agric Biol Eng 2016;9(5):143–8.

[35] Zhang F, Wu BF, Luo ZM. Winter wheat yield predicting for America using remote sensing data. J Remote Sens 2004;8:611–7.

[36] Du M, Noguchi N. Monitoring of wheat growth status and mapping of wheat yield's within-field spatial variations using color images acquired from UAV-camera system. Rem Sens 2017;9(3):289.

[37] Khan Z, Chopin J, Cai J, Eichi V-R, Haefele S, Miklavcic SJ. Quantitative estimation of wheat phenotyping traits using ground and aerial imagery. Rem Sens 2018;10(6):950.

[38] Jin X, Liu S, Baret F, Hemerlé M, Comar A. Estimates of plant density of wheat crops at emergence from very low altitude UAV imagery. Remote Sens Environ 2017;198: 105–14.

[39] Banu TP, Borlea GF, Banu C. The use of drones in forestry. J Environ Sci Eng B 2016; 5(11):557–62.

[40] Tack F, Merlaud A, Meier AC, Vlemmix T, Ruhtz T, Iordache MD, et al. Intercomparison of four airborne imaging DOAS systems for tropospheric NO2 mapping—the AROMAPEX campaign. Atmos Meas Tech 2019;12(1):211–36.

[41] Marchica A, Loré S, Cotrozzi L, Lorenzini G, Nali C, Pellegrini E, et al. Early detection of sage (*Salvia officinalis* L.) responses to ozone using reflectance spectroscopy. Plants 2019;8(9):346.

[42] Pajares G. Overview and current status of remote sensing applications based on unmanned aerial vehicles (UAVs). Photogramm Eng Rem Sens 2015;81(4):281–330.

[43] Lan Y, Chen S. Current status and trends of plant protection UAV and its spraying technology in China. Int J Precision Agric Aviation 2018;1(1):1–9.

[44] Manfreda S, McCabe MF, Miller PE, Lucas R, Pajuelo Madrigal V, Mallinis G, et al. On the use of unmanned aerial systems for environmental monitoring. Rem Sens 2018; 10(4):641.

[45] Tripicchio P, Satler M, Dabisias G, Ruffaldi E, Avizzano CA, editors. Towards smart farming and sustainable agriculture with drones. International conference on intelligent environments; 2015 15−17 July 2015.

[46] Mink R, Dutta A, Peteinatos GG, Sökefeld M, Engels JJ, Hahn M, et al. Multi-temporal site-specific weed control of *Cirsium arvense* (L.) scop. And *Rumex crispus* L. in maize and sugar beet using unmanned aerial vehicle based mapping. Agriculture 2018;8(5): 65.

[47] Lawes R, Chen C, Whish J, Meier E, Ouzman J, Gobbett D, et al. Applying more nitrogen is not always sufficient to address dryland wheat yield gaps in Australia. Field Crop Res 2021;262:108033.

[48] Singh RK, Berkvens R, Weyn M. AgriFusion: an architecture for IoT and emerging technologies based on a precision agriculture survey. IEEE Access 2021;9:136253−83.

[49] Kerkech M, Hafiane A, Canals R. Vine disease detection in UAV multispectral images using optimized image registration and deep learning segmentation approach. Comput Electron Agric 2020;174:105446.

[50] Su J, Liu C, Coombes M, Hu X, Wang C, Xu X, et al. Wheat yellow rust monitoring by learning from multispectral UAV aerial imagery. Comput Electron Agric 2018;155: 157−66.

CHAPTER 5

Applications of geospatial technologies for precision agriculture

Mobushir R. Khan[1], Richard A. Crabbe[1], Naeem A. Malik[2], Lachlan O'Meara[1]

[1]*School of Agricultural, Environmental and Veterinary Sciences, Charles Sturt University, Albury, NSW, Australia;* [2]*Institute of Geo-Information & Earth Observation, PMAS-Arid Agriculture University Rawalpindi, Rawalpindi, Punjab, Pakistan*

5.1 Introduction

Advances in the field of science and technology have greatly improved the global food security situation. Norman E. Borlaug paved the way to "green revolution" by introducing high-yielding and short-statured disease- and lodging-resistant wheat. Since, then researchers have strived to improve the production of food, fiber, and fodder crops in the fields. However, increased food production has certain negative impacts on the environment owing to heavy use of chemicals [1]. These chemicals are used as fertilizers, herbicides, and insecticides to suppress yield limiting soil constraints, weeds, and insects, respectively. Further, additional land was cleared of other uses such as forests and wetlands to create more farms for food production. The situation was further exacerbated when we started to use more water for production including surface water as well as groundwater. This has deteriorated the groundwater quantity and quality. The world population is expected to exceed 9.7 billion by 2050 from today's 8 billion [2]. It will continue to exert pressure on scarce natural resources including both the land and water. As a result, fresh water and arable land are also getting depleted and degraded, respectively. Therefore, agriculture remains the most important concern for sustainable livelihood [3]. Precision Agriculture (PA) offers a sustainable management solution to meet increasingly high demand of food and raw materials associated with it while keeping environment safe.

In 1997, the US House of Representatives defined the term PA as farming management philosophy based on integrated information system with the objective to enhance the whole farm's profitability with optimum inputs and minimum harmful impacts on ecosystems and environment. Among many definitions, PA was simply defined as "the application of right input at right location and at right time" [4].

PA minimizes the climate change effects through spot applications of chemicals, water conservation while improving the microclimate. Therefore, it helps in achieving sustainable agriculture for improving food security and hence becomes a linchpin of Climate Smart Agriculture [5]. Among other benefits, PA has also proved to be an economical and profitable approach in 68% of the cases studied [6]. Moreover, it is argued that the goal of sustainable agriculture can be achieved through the adoption of PA by increasing the ability to interpret farms' big data and generate knowledge to improve the crop produce [7].

5.2 Management philosophy of PA

The management philosophy of PA is based on identifying and managing, within field and seasonal, variabilities of crop limiting factors including soil, weather, water, crop genetic potential, and management operations. Moreover, the variations of the factors must be structured, of sufficient magnitude, and manageable. It is advised to diagnose and remove the causes of variations and apply uniform input applications. However, variable rate treatments are recommended if the causes of variation cannot be eliminated. This philosophy can be applied as sequence of operations covering the complete paddock to plate chain.

The main aim of PA is to apply inputs as per variations in crop and soil requirements which was not possible until the development of variable rate technologies in 1980s. PA is referred as a management philosophy and its concepts and objectives were only achievable with the advent of its enabling technologies.

5.3 Enabling technologies of PA

As agriculture has become information extensive, farmers must overcome the challenges of poor access to knowledge, financial services, markets, and ancillary information flows. Farmers need information such as which crop is most suitable for his land and how to grow that crop, amount of water required, which part of his land is performing better, and where to sell his produce and where to buy the required products [8]. For example, precise information on soil moisture and weather can help farmers in deciding when to sow and irrigate their crops. Farm and field level data can be collected with ease using location-based and remote sensors. These can then be integrated through the Internet of Things (IoT) to make agriculture a digital entity. Similarly, remotely sensed imagery can be used for yield prediction and stress monitoring. Early crop stress detection and monitoring are essential for achieving higher yields and hence increasing productivity. The use of smart systems can significantly improve the quality and efficiency of precision agriculture by recognizing stressed area patterns from real-time (satellite and drone) imagery and by real-time data-driven farm operations management (tillage, irrigation, pesticides, and fertilizers applications) [9]. The technologies which allow the implementation of PA management philosophy are discussed below.

5.3.1 Geographic Information Systems

Development and availability of positioning technologies like Global Positioning System (GPS) and Global Navigation Satellite System (GNSS) for civilian uses particularly agriculture have paved the way for large-scale adoption of site-specific management. These technologies led to development of systems such as auto steering, controlled traffic farming, yield monitors, and variable rate technologies (VRT). VRTs are comprised of differential global positioning system (DGPS), processing unit, variable rate software, and controllers. These are designed for optimized seeding, planting density, and application of fertilizers, weedicide/herbicides, and insect/pesticides to reduce inputs and environment risks. It was reported by Neupane (2019) that variable rate irrigation reduces water use by between 10% and 15% [10].

Similarly, with the fabrication of sensors for mapping in situ and on-the-go soil and plant characteristics has led to the development of useful products like yield maps, fertilizer prescription maps. These help in developing optimal management zones at field level. These products provide deep insights for better field management activities as compared to traditional ways of farming. For example, sensors for mapping and monitoring of apparent electrical conductivity, soil moisture, and microclimate of plants were readily available.

5.3.2 Remote sensing (RS)

Crops undergo biophysical and biochemical changes while reacting to biotic and abiotic stresses. It is important to know when such changes occur, and it necessitates the use of remote sensing in agriculture. Various RS systems ranging from gamma rays to microwave have been used to detect infection symptoms, physiological responses, and structural changes in plants [11–13]. These RS systems for plant diseases were categorized by Zhang et al. [14] into four categories, that is, visible and near-infrared (Vis-NIR) spectral systems, fluorescence and thermal systems, synthetic aperture radar (SAR), and light detection and ranging (LiDAR).

Airborne and space-borne sensors provide georeferenced information on crop health while recording plant spectral profiles with varying characteristics depending upon the wavelengths at which data is recorded. Among these, sensors mounted on unmanned aerial vehicles (UAVs) can provide high spectral, spatial, and temporal resolution imageries for mapping and monitoring of agricultural systems and processes [15]. It was reported that remote sensors were the most used technology in precision agriculture [16]. Remote sensors can be classified based on the platform or the source of energy. Platforms can be space, aerial, or ground-based. Both active and passive remote sensors are being used in precision agriculture for mapping and monitoring plant biotic and abiotic stresses. Data resolution required for PA is defined by the management objectives, size of the field, and equipment to be used for variable rate application [17].

74 CHAPTER 5 Applications of geospatial technologies

Imagery from high spatial resolution satellites such as IKONOS with 1—4 m was found adequate for mapping of soils, prediction of crop yields, nutrient management, and evapotranspiration (ET) estimation [18—21]. Data from Pleiades-1A and Worldview-3 having spatial resolution less than 2 meters (2 m) can be used for detection of disease and water stress [22—24]. Availability and flexibility of UAVs in acquiring very high spatial resolution (centimeter level) imagery has certainly paved the way for precision agriculture's development. Fig. 5.1 presents optimal ranges of the spatial and temporal resolution along with spectral resolution (can be accessed through the platform) needed for precision agriculture and allied disciplines.

Sophisticated procedures have been and are being developed for transforming georeferenced data recorded by sensors into information that can be used in the decision-making process. Accuracies of these procedures depend upon image resolution, atmospheric and crop conditions along with analytical techniques being employed, that is, deep learning, regression-based or simulations [17].

Utilization of remote sensing datasets in PA is limited by huge volume of data as high spectral, spatial, radiometric, and temporal resolution data are required for site-specific management [26]. Advanced computational techniques for processing

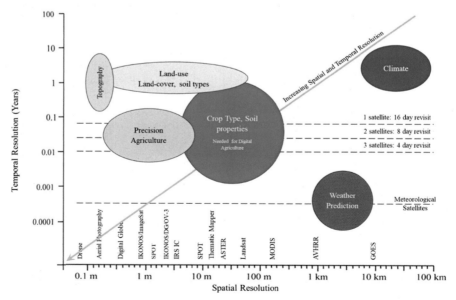

FIGURE 5.1

Spatiotemporal characteristics of RS-based system for precision agriculture.

Modified after Jensen JR. Introductory digital image processing. A remote sensing perspective. Pearson series in geographic information science; 2015.

remotely sensed big data (huge volume, variety, and velocity of data) are being developed to use crucial information in decision support systems for PA.

PA technologies based on remotely sensed imagery, for example, variable rate fertilizer application in green seeker and crop circles have been already adopted for commercial agriculture [17]. Plant water stress can be monitored by utilizing thermal sensors for estimation of temperatures [27]. Similarly, microwave sensors are very useful in estimating soil moisture content and water consumption by crops over large areas [28]. Although data from microwave sensors is of course resolution but it can be downscaled for PA use.

5.3.3 Big data analytics

Over the last decade, artificial intelligence has been applied in agriculture owing to its capacity to identify complex patterns in the big datasets. The potential of machine learning techniques to monitor crop stresses using digital imagery have been explored to a great extent during past 10 years.

Despite the exponential development of remote sensing technologies, the commercial use of remotely sensed imageries in PA is limited by lack of readily available imaging solutions, high budgetary requirements, cloud cover limitations, and automated systems for processing and analysis of imageries [29]. For wider adoption of remote sensing data in PA, more friendly and accurate systems are required [17].

Moreover, more case studies have been carried out to link crop limiting factors with appropriate management actions like sowing, irrigation, fertigation, harvesting, and marketing. Common imaging techniques used for detecting plant stress include digital fluorescence, LiDAR, thermography, multispectral and hyperspectral techniques [30].

Data analytic capabilities provide knowledge and through Information and Communication Technologies (ICTs) this knowledge is transferred to the farmers, thus, enabling them to practice sustainable agriculture by decreasing inefficiencies, costs, and enhancing farm profitability [31]. Web services are introduced to transfer the knowledge through development of decision support systems. Digital technologies possess huge potential in meeting the demand for food and play their role throughout the value chain [32].

During the past decade, the potential of advanced techniques like machine learning, internet of things (IoTs), and blockchain had been explored to a great extent. These techniques have had significant impact in the agriculture sector with applications like identification of weeds through artificial intelligence, assessment of fruit ripening, and identification of nutrient deficiencies. IoTs have already contributed significantly in modernized agriculture [33].

Every day new aspects and the application of AI techniques in agriculture are being reported. The potential of machine learning techniques to monitor crop stresses using digital imagery have been explored to a large extent in the past 10 years [34]. Generally, deep learning has proved to be promising in plant stress detection through image classification, segmentation, and object detection, which can be instrumental in development of precision agriculture [35]. Similarly, transfer deep learning can be used for development of disease heat maps, which can be used for early detection and incubation period evaluation [36].

5.4 State-of-the-art conceptual system

Examples of PA solutions currently being adopted in developed countries and likely to be adopted in developing countries are given below.

5.4.1 Soil management system

Soil management systems are capable of characterizing soil variabilities in terms of their physical and chemical properties including slope and cultivation history. This system provides the basis of variable rate irrigation, fertigation and seeding. Soil quality is maintained through efficient monitoring of soils. Farming communities need to be capable of monitoring the status of their soil as vital indicator of the health of their land. Fig. 5.2 shows a mobile application which is a component of web-based decision support system that enables farmers to access soil properties along with agro-meteorological data. The system was developed under Australian

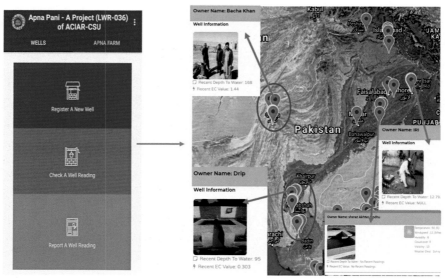

FIGURE 5.2

Mobile-based decision support system for optimal management of water resources at the farm level.

Note: Application is developed for "Improving groundwater management to enhance agriculture and farming livelihoods in Pakistan" a project funded by Australian Center for International Agricultural Research (ACIAR).

Center for International Agricultural Research (ACIAR) funded projects implemented at various locations of Indus Basin, Pakistan [37]. The system also assists farmers in computing crop water requirements, provides information on the soil status of their land, and crop profitability. Thus, it enables them to select the best crop to grow among various options.

5.4.2 Auto steering system

Such systems allow machines to follow certain tracts and hence enhancing accuracies and reducing operator-induced errors, inputs, and fatigue. Autosteering systems are being used for planting, harvesting, and agronomic operations. These systems are installed on tractors and/or farming machinery and are paired with precise GPS to provide pass to pass accuracy in operations.

5.4.3 Precise seeding system

Precise seeding systems are applied to sow maximal germination percentage and minimize inputs while ensuring optimum plant densities. Currently collaborating farmers in the aforementioned project is supported through satellite imagery for crop husbandry. In Fig. 5.3 Soil-Adjusted Vegetation Index is presented showing areas with less germination in red as compared to green areas. Sentinnel-2 at 10 m resolution data is used to extract this information as per following equation.

$$MSAVI = \frac{2nir + 1 - \sqrt{(2NIR + 1)^2 - 8(NIR - RED)}}{2}$$

Additionally, maps were built based on this index during the early vegetative state to guide variable rate fertilizer application.

5.4.4 Variable rate fertilizer application

This system utilizes information from maps of soil properties, management zones, and crop nutrient requirements to apply fertilizers in the field. Consequently, farm input costs are reduced while keeping the environment safer.

5.4.5 Variable rate irrigation

This system increases irrigation efficiencies based on sensor's real-time input. As a result, land and water productivity is increased.

5.4.6 Variable rate sprayer

Crop-specific AI-based algorithms are developed to apply variable rate herbicides, weedicides, and pesticides to minimize inputs and reduce emissions of greenhouse gases while producing quality produce.

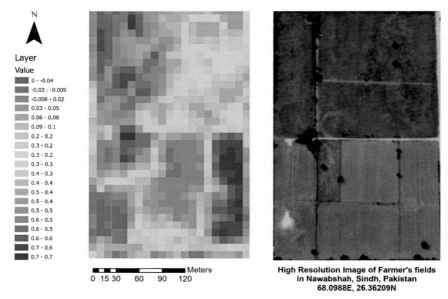

FIGURE 5.3

Modified soil-adjusted vegetation index (MSAVI) of a farmer's field.

5.4.7 Crop stress monitoring system

Various types of crop stress monitoring systems are being deployed. These include deep learning–based image processing systems for crop health monitoring through identification of water, pest, weed stresses.

To monitor stress, knowledge of chlorophyll content in the crop is necessary, as it is an indicator of crop photosynthesis. In the aforementioned field, red-edge band of Sentinnel-2 was used to prepare a Chlorophyll index (Red-edge Chlorophyll Index). It shows the photosynthetic activity of the crop (Fig. 5.4)

5.4.8 Yield monitoring system

Yield monitors allow recording of yield on the go while crop is being harvested through grain flow sensors and GNSS. Information from yield monitors is used to develop crop yield maps which are then correlated with crop limiting factors to diagnose the causes of low performing portions of the field.

In a nutshell, Fig. 5.5 is the conceptual diagram which highlights the use of various enabling technologies for precision agriculture through the agricultural value chain, that is, from paddock to plate.

5.4 State-of-the-art conceptual system   79

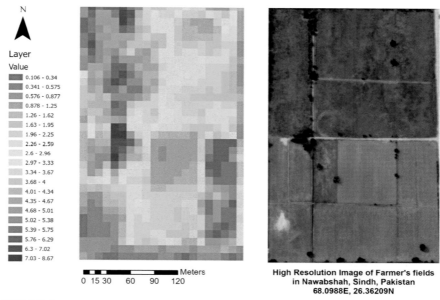

FIGURE 5.4

Red-edge chlorophyll index of a farmer's field.

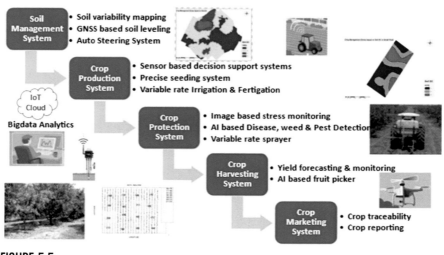

FIGURE 5.5

State-of-the-art conceptual model of enabling technologies of PA.

5.5 Perceptions of digital technologies

Widespread adoption of PA is subjected to the development of creative policies by the stake holders and development of incentives based and cost-sharing mechanisms. An effort was made to assess the current and future state of precision agriculture in Australia. A survey was undertaken to investigate the perceptions of various Australian agricultural industry stakeholders and their perceptions of digital technologies in agriculture and how this will affect the stakeholder groups and sustainability (Fig. 5.6). Participants were also asked what technologies they would use or avoid, and if and how each could be used to increase sustainability.

The results of the survey showed interest of stakeholders in adopting the technologies which form the core of PA. The technologies include (in the order of preference) drones, GPS, satellite remote sensing, and automation.

The respondents also believed that the biggest barriers to adoptions are financial cost and knowledge of the products as shown by Fig. 5.7.

This demands actions from the government (relevant) agencies, research organizations and academia to promote such technologies and create government-academia-industry linkages to promote the use of such technologies in a sustainable manner.

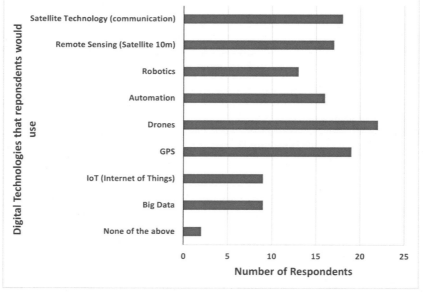

FIGURE 5.6

Stakeholders' preferred technologies to be adopted.

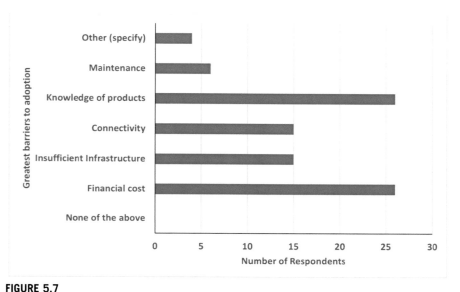

FIGURE 5.7

Stakeholders' perception of barriers to adoption of technologies in PA.

5.6 Conclusions

Precision Agriculture (PA) offers a sustainable management solution to meet the increasingly high demand for food and fiber while keeping the environment safe. Crop management (from sowing to harvesting) is an important component of the agricultural value chain and needs apt decisions by the farmers. This whole process is dynamic in nature and geospatial technologies assist in identifying the dynamics such as variabilities in soil, weather, water, and crop performance. Such technologies in combination with IoT guide farmers not only to identify these yield-limiting factors but also to provide support in making optimal decisions. This chapter describes a state-of-the-art conceptual system of PA while presenting practical examples of using geospatial technologies for PA. The development in the field of remote sensing and GIS has enabled us to acquire field-scale data and convert that information into knowledge. Now the farmers and extension workers can be enabled to acquire that knowledge without the need of running complex algorithms and acquire data as that is done through big data analytics in a cloud environment. When it comes to using digital technologies, agricultural stakeholders prefer using drones, GPS, and satellite-based remote sensing data. However, for large-scale adoption of PA-enabling technologies, we need to reduce the cost of such technologies (through government support or cooperative farming) and improve the infrastructure, especially network connectivity. We conclude that using geospatial technologies in combination with IoT and cloud computing not only enhances a paddock's productivity but also contributes to environmental sustainability.

References

[1] Kleijn D, Rundlöf M, Scheper J, et al. Does conservation on farmland contribute to halting the biodiversity decline? Trends Ecol Evol 2011;26(9):474–81.

[2] World population prospects 2019. United Nations: Department of Economic and Social Affairs PD; 2019. Available at: https://population.un.org/wpp/. [Accessed 5 August 2022].

[3] Trendov N, Varas S, Zeng M. Digital technologies in agriculture and rural areas. Rome: Food and Agriculture Organization of the United Nations; 2021.

[4] Gebbers R, Adamchuk VI. Precision agriculture and food security. Science 2010;327(5967):828–31.

[5] Food and Agriculture Organization (FAO) of the United Nations. Facing the challenges of climate change and food security, ISBN 978-92-5-107737-5.

[6] Griffin T, Lambert D, Lowenberg-DeBoer J. Testing appropriate on-farm trial designs and statistical methods for precision farming: a simulation approach. In: Proceedings of the 7th international conference on precision agriculture and other precision resources management. vols. 25–28. Minneapolis, MN, USA: Hyatt Regency; 2004. p. 733–1748.

[7] Kamble SS, Gunasekaran A, Sharma R. Modeling the blockchain enabled traceability in agriculture supply chain. Int J Inf Manag 2020;52:309–31.

[8] EST GR, Sylvester G. Information and communication technology (ICT) in agriculture—a report to the G20 agricultural deputies. 2017.

[9] Budaev D, Lada A, Simonova E, et al. Conceptual design of smart farming solution for precise agriculture. Manag Appl Complex Syst 2019;13:309–16.

[10] Neupane J, Guo W. Agronomic basis and strategies for precision water management: a review. Agronomy 2019;9(2):87.

[11] Hahn F. Actual pathogen detection: sensors and algorithms—a review. Algorithms 2009;2(1):301–38.

[12] Mahlein A. Precision agriculture and plant phenotyping are information-and technology-based domains with specific demands and challenges for. Plant Dis 2016;100:241–51.

[13] Sankaran S, Mishra A, Ehsani R, et al. A review of advanced techniques for detecting plant diseases. Comput Electron Agric 2010;72(1):1–13.

[14] Zhang J, Huang Y, Pu R, et al. Monitoring plant diseases and pests through remote sensing technology: a review. Comput Electron Agric 2019:165.

[15] Maes WH, Steppe K. Perspectives for remote sensing with unmanned aerial vehicles in precision agriculture. Trends Plant Sci 2019;24(2):152–64.

[16] Cisternas I, Velásquez I, Caro A, et al. Systematic literature review of implementations of precision agriculture. Comput Electron Agric 2020:176.

[17] Sishodia RP, Ray RL, Singh SK. Applications of remote sensing in precision agriculture: a review. Rem Sens 2020;12(19):31–6.

[18] Seelan SK, Laguette S, Casady GM, et al. Remote sensing applications for precision agriculture: a learning community approach. Remote Sens Environ 2003;88(1–2):157–69.

[19] Enclona E, Thenkabail P, Celis D, et al. Within-field wheat yield prediction from IKONOS data: a new matrix approach. Int J Rem Sens 2004;25(2):377–88.

[20] Sullivan DG, Shaw J, Rickman D. IKONOS imagery to estimate surface soil property variability in two Alabama physiographies. Soil Sci Soc Am J 2005;69(6):1789−98.
[21] Yang G, Pu R, Zhao C, et al. Estimating high spatiotemporal resolution evapotranspiration over a winter wheat field using an IKONOS image based complementary relationship and Lysimeter observations. Agric Water Manag 2014;133:34−43.
[22] Navrozidis I, Alexandridis TK, Dimitrakos A, et al. Identification of purple spot disease on asparagus crops across spatial and spectral scales. Comput Electron Agric 2018;148:322−9.
[23] Salgadoe ASA, Robson AJ, Lamb DW, et al. Quantifying the severity of phytophthora root rot disease in avocado trees using image analysis. Rem Sens 2018;10(2):226.
[24] Bannari A, Mohamed AM, El-Battay A. Water stress detection as an indicator of red palm weevil attack using worldview-3 data. In: 2017 IEEE international geoscience and remote sensing symposium (IGARSS). IEEE; 2017. p. 4000−3.
[25] Jensen JR. Introductory digital image processing. A remote sensing perspective. Pearson series in geographic information science; 2015.
[26] Huang Y, Chen Z-x, Tao Y, et al. Agricultural remote sensing big data: management and applications. J Integr Agric 2018;7(9):1915−31.
[27] Khanal S, Fulton J, Shearer S. An overview of current and potential applications of thermal remote sensing in precision agriculture. Comput Electron Agric 2017;139:22−32.
[28] Palazzi V, Bonafoni S, Alimenti F, et al. Feeding the world with microwaves: how remote and wireless sensing can help precision agriculture. IEEE Microw Mag 2019;20(12):72−86.
[29] Katsigiannis P, Galanis G, Dimitrakos A, et al. Fusion of spatio-temporal UAV and proximal sensing data for an agricultural decision support system. In: Fourth international conference on remote sensing and geoinformation of the environment (RSCy2016). SPIE; 2016. p. 564−74.
[30] Bradshaw JE. Review and analysis of limitations in ways to improve conventional potato breeding. Potato Res 2017;60(2):171−93.
[31] El Bilali H, Allahyari MS. Transition towards sustainability in agriculture and food systems: role of information and communication technologies. Inf Process Agric 2018;5(4):456−64.
[32] USAID. Digital tools in USAID agricultural programming toolkit. 2018.
[33] Khanna A, Kaur S. Evolution of internet of things (IoT) and its significant impact in the field of precision agriculture. Comput Electron Agric 2019;157:218−31.
[34] Barbedo JGA. Detection of nutrition deficiencies in plants using proximal images and machine learning: a review. Comput Electron Agric 2019;162:482−92.
[35] Gao Z, Luo Z, Zhang W, et al. Deep learning application in plant stress imaging: a review. AgriEngineering 2020;2(3):29.
[36] Wu H, Wiesner-Hanks T, Stewart EL, et al. Autonomous detection of plant disease symptoms directly from aerial imagery. Plant Phenome J 2019;2(1):1−9.
[37] Talbot S. Adapting to salinity in the southern Indus Basin. 2021. Available at: https://www.csu.edu.au/research/ilws/research/summaries/2021/adapting-to-salinity-in-the-southern-indus-basin.

CHAPTER 6

Precision irrigation: challenges and opportunities

Muhammad Naveed Anjum[1,2], Muhammad Jehanzeb Masud Cheema[1,3], Fiaz Hussain[1], Ray-Shyan Wu[4]

[1]*Department of Land and Water Conservation Engineering, Faculty of Agricultural Engineering and Technology, PMAS-Arid Agriculture University Rawalpindi, Rawalpindi, Punjab, Pakistan;* [2]*Data Driven Smart Decision Platform, PMAS-Arid Agriculture University Rawalpindi, Rawalpindi, Punjab, Pakistan;* [3]*National Center of Industrial Biotechnology, PMAS-Arid Agriculture University Rawalpindi, Rawalpindi, Punjab, Pakistan;* [4]*Department of Civil Engineering, National Central University, Chung-Li, Taiwan*

6.1 Introduction

Agriculture plays an indispensable role in economic growth, food security, employment generation, and poverty alleviation. For Pakistan, the agricultural sector contributes 19.2% to GDP (Gross Domestic Product) and employs 38.5% of the country's workforce (Pakistan Economic Survey 2020−21). The traditional methods (such as surface irrigation) of irrigation that the farmers used are not sufficient to fulfill the increasing demand of food. Therefore, the adaptation of sophisticated precision irrigation technologies such as drones, soil moisture and temperature sensors, crop sensors, internet of things (IoT), robots, etc., is a way forward to help farmers' challenges posed by population growth, water scarcity, and climate change. Agriculture experts believe that water sense technology can reduce an extra 20% of irrigation water use along with modern irrigation techniques (drip and sprinkler). Moreover, smart soil moisture sensors can measure the moisture and water level in soil and transmit data updates to a cloud on regular intervals where farmers receive information. In order to avoid under or over-irrigation of the crop, the farmer receives continuous information about how much and where to irrigate [1].

Artificial Intelligence (AI) in agriculture has enhanced crop production, saved the excess use of water, pesticides, herbicides, and improved real-time monitoring, harvesting, processing, and marketing [2]. Automatic irrigation scheduling techniques replaced the manual irrigation which was based on soil water measurement. An automated irrigation monitoring system consists of wireless sensor network (WSN), Internet of Things (IoT), and Machine Learning (ML) algorithms, which solve the problems of over- and under-irrigation based on crops and weather scenarios [3]. The technology of smart irrigation uses soil moisture sensors to measure the soil moisture content accurately and determine the moisture level and transmit this reading to the controller for irrigation [4]. When the scheduled time arrives, the sensor reads the moisture content or level for that particular zone, and watering will be allowed in that zone only if the moisture content is below the threshold [5]. Automation of irrigation systems can lead to increased crop productivity and water saving [6,7]. There are considerable advantages of automated irrigation systems over traditional irrigation systems. Savitha and Maheshwari [8] documented that the automated irrigation system in India had increased the crop production by up to 40%. Deekshithulu et al. [9] found that, as compared with the flood irrigation system, about 36% of fresh water was saved using automated irrigation system. A study by Al-Ghobari et al. [10] found that using an automated drip irrigation system in Saudi Arabia reduced water consumption by up to 26% while increasing crop productivity.

Precision agriculture is hampered by the high cost of sensors and other latest technologies/equipments. In order to facilitate widespread use of smart technologies/equipments by farmers, low-cost systems must be made available for them. Farming with precision necessitates the adoption of certain technologies, so growers/farmers can have confidence in the appliance's capabilities even if they don't know all of the specifics [11]. In Pakistan, farmers hold small farms and annual income below standards while they are less aware of innovative technologies. Moreover, the initial cost of such a system is high. Common barriers are influencing the adoption of technological innovations [12]; others include (1) farm size, (2) risk exposure and capacity to bear risks, (3) human capital, (4) labor availability, (5) credit constraint, (6) tenure, and (7) access to commodity markets [13].

Irrigation is one of the primary sources of freshwater consumption in most countries, and globally, its efficient use and management is a leading concern. Surprisingly, agriculture accounts for over 90% of all freshwater consumption in Pakistan, and nearly 50% is wasted due to poor irrigation practices. Therefore, this book chapter covers information about traditional irrigation methods and precision irrigation (an automated monitoring system), its challenges and opportunities for improving irrigation water-use efficiency. Water-use efficiency (WUE: crop yield per unit of water use) or water productivity (WP) is the concept used to indicate the efficiency of the crop production system relative to the amount of applied irrigation. With the increasing scarcity of available water, the WUE enhancement is the main criteria to overcome controversy about using water resources by agriculture [14]. Because of the water security problem, Pakistan is potentially at the risk of experiencing a severe food crisis in the near future. According to the forecast of World Bank's 2020–21 report, water shortages are expected to rise to 32% by 2025, resulting in a food scarcity of about 70 million metric tonnes [15]. Due to the country's water scarcity, proper water usage is essential for the provision of safe drinking water as well as long-term agricultural and industrial development. Agriculture, the backbone of the country's economy and the primary source of food, however, is extremely vulnerable to fluctuations in water supply. In the wake of an imminent water crisis, inclusive and comprehensive planning is imperative (Pakistan Economic Survey 2020–21, Chapter 2 Agriculture).

Higher water productivity can be achieved by increasing production per unit of water consumed, or by maintaining the same production while reducing water use [16,17]. There is a lack of water-use efficiency (WUE) awareness in Pakistan [18]. Irrigation efficiency typically falls between 38.7% and 42.6%. Indian farmers produce 1 kg/m^3, whereas the Californians are able to produce 1.5 kg/m^3, and the production in Pakistan is just 0.5 kg/m^3. There is a big gap between the potential and the actual yield level [19]. Lack of irrigation water, inefficient irrigation practices, inadequate irrigation scheduling, and incorrect conjunctive use of surface and groundwater are the primary causes of low WUE in the agricultural sector. Innovative agronomic and water management approaches could enhance WUE, but this would require changes in agricultural practices and policies. With Precision Irrigation Systems, water shortage would no longer be a serious issue. At the farm, high efficiency irrigation systems (sprinkler and drip) and properly lined channels might result in an irrigation efficiency of 81.2% [13].

Irrigated agriculture has always played an important role in the economic development of Pakistan. A total of 23.4 million hectares are cultivated, with the majority of the land being farmed by small landowners, who have an average farm size of 6.4 acres. Irrigated land accounts for approximately 90% of all cultivatable agricultural land and is crucial for agriculture's overall production. Productivity in the agricultural sector is decreasing, which is affecting economic growth and raising questions about food security. In this scenario, Pakistan has to adopt precision agriculture technology and innovation in order to conserve and enhance the country's natural resources while also increasing productivity. A 70% increase in agricultural productivity by 2050 is predicted by the Food and Agriculture Organization (FAO) of the United Nations based on adoption of IoT-based agricultural practices. Farmers in Pakistan's rural areas can make better use of their limited inputs and resources by implementing this technique on a larger scale across the country. Unfortunately, Pakistan has lagged behind other countries in adopting modern agricultural technologies. In contrast to other emerging nations that have embraced agricultural innovation, we are still firmly stuck in the traditional methods. We must utilize our natural resources in a way that allows farmers to produce more with less inputs. Precision agriculture and higher crop yield can both benefit from the promotion of cutting-edge technologies like precision irrigation.

6.2 Irrigation application methods

Irrigation is the application of water to soils for crop use. Irrigation water is applied to the fields by following different methods.

1. Surface irrigation methods
    (a) Basin irrigation.
    (b) Border irrigation.
    (c) Furrow irrigation.

2. Pressurized irrigation method
   (a) Drip irrigation.
   (b) Sprinkler irrigation.

Surface irrigation methods are traditional methods and referred as flood irrigation or gravity irrigation methods since gravity is the driving force causing water flow. The objectives of any surface irrigation method are to apply efficiently just enough quantity of water to satisfy crop water requirements; to ensure even distribution of water on the entire field; and to minimize the losses of runoff and/or deep percolation. To accomplish these objectives, the required water depth should be provided at each irrigation, which depends upon water availability, field level, slope, soil characteristics, crop type, and cultural practices.

Micro-irrigation systems, including drip, surface/subsurface, and sprinkler irrigation methods, are under the category of precision water management technologies. These systems are sometimes referred to as High Efficiency Irrigation Systems (HEISs). There are many different types of HEISs, including drip, bubblers, sprinklers, rain-guns, center pivots, etc. The application efficiency of surface and pressurized irrigation methods depends on their design, management, and operation. According to literature studies, the application efficiency of various irrigation methods was estimated: basin irrigation (up to 40%); border irrigation (40%−60%); furrow irrigation (50%−70%); sprinkler irrigation (50%−90%); and drip irrigation (65%−95%). Traditional irrigation methods have an efficiency of 40%−70%, while drip and sprinkler irrigation systems have an efficiency of up to 95%. Pressurized irrigation systems can save 35%−65% water compared with traditional irrigation systems.

A high-performing irrigation system must be able to irrigate uniformly and at the required depth at the appropriate time. For this reason, irrigation scheduling is the most commonly used tool for determining "when" and "how much" water should be applied to a crop. In order to ensure ideal soil water conditions for crop production, water management decisions have a significant impact on how uniform water may be applied via various irrigation methods. Irrigation systems that are properly designed, constructed, maintained, and managed save energy and money by reducing the amount of applied irrigation water.

The definition of various irrigation methods are outlined below:

Gravity irrigation: A technique in which water is distributed in the field by gravity.

Flood irrigation: A method in which the entire field is ponded with flooded water.

Basin irrigation: In this method, the water is applied to leveled field that is bounded by dikes/ridges or check banks to stop runoff.

Border irrigation: Method of irrigation in which parallel border strips are used where the water flows down the slope at a nearly uniform depth.

Furrow irrigation: In this method water is applied to crop field using furrows (small parallel channels) to favor water infiltration while advancing down the field.

Pressurized irrigation: Irrigation system in which water is pumped and flows to the crop field by pressure.

Sprinkler irrigation: Sprinkler irrigation is an overhead irrigation system that sprays water on the land or crop in a manner similar to rain.

Drip irrigation or trickle irrigation: An irrigation system in which water is applied directly to the root zone of plants by means of small emitters in the form of droplets.

Bubbler irrigation: Similar to trickle irrigation, this technique uses microsprinklers that are fixed on small spikes to deliver water to the plants.

All irrigation methods have their advantages and disadvantages and the farmer must know which irrigation method suits the local conditions best. The best way to make an informed decision about irrigation systems is to test and simulate (for surface irrigation: SURDEV, SIRMOD, WinSRFR, SADREG, COBASIM, BASCAD, etc.) them in the context of the local environment.

6.2.1 Surface irrigation methods

Surface irrigation is the traditional water application method in the world. About 84% of the world uses the surface irrigation method to irrigate the crops [20,21]. The flow is started on one side of the field and progressively spread out until the entire field is irrigated. This can be accomplished by flooding the entire field at once (basin irrigation), putting water into small channels (furrows), or flooding the land surface strips (borders) until the entire field is covered. Fig. 6.1 is used to explain the surface irrigation methods (basin, border, and furrow).

In reality, most of the Pakistani farmers employ these traditional methods to irrigate their fields. They apply water at the inlet, which flows downstream. As it moves over the field, some of the water seeps into the ground. The primary design objectives in surface irrigation are to maximize application efficiency and uniformity with primary design variables (inlet flow rate and application time). Slope and soil infiltration rate are crucial for successful functioning of surface irrigation systems. It is possible that water will infiltrate deeper into the soil near to the inlet than downstream if the soil infiltration rate is high. Having steep slopes in the field can lead to excessive water runoff, erosion, and the formation of depressions on the ground. Slopes of less than 2% are generally considered acceptable. One of the most significant sources of water waste for irrigation systems is runoff from fields at the downstream end. In furrow irrigation, 20%–30% of the applied water may need to run off the field in order to achieve somewhat consistent water distribution. In the absence of a runoff recovery or reuse system, application efficiency can go as low as 60%–70%.

6.2.2 Pressurized irrigation method—sprinkler irrigation

Sprinkler irrigation system is a pressurized (overhead) irrigation where water is generally applied over the crop canopy in an identical way to rainfall. Following

FIGURE 6.1

Surface irrigation methods (A) basin irrigation, (B) border irrigation, and (C) furrow irrigation.

are the primary distinctions between conventional surface irrigation and pressurized irrigation methods.

- A significant difference between flow regimes of water: Conventional surface irrigation methods require a large stream size; on the other hand pressurized irrigation systems can use minimal flows, as low as 1 m^3/h.
- The flow direction along the route: In conventional surface methods, water distribution follows field contours by the force of gravity as overland flow; while in pressurized irrigation system, water is distributed in closed pipes under pressure along possible shortest route, regardless of the terrain slope and contour.
- The total field area is irrigated simultaneously: Conventional surface methods apply large volumes of water per unit of area, whereas pressurized irrigation systems spread small amounts of water over a vast area at a time.

- The conventional surface gravity methods do not require the use of external energy, piped irrigation systems need an external source of energy in the form of pressure, which is supplied by a pumping unit or a supply tank located at a high point.

Sprinkler systems typically include an electric or diesel pumping unit, a main pipeline with predetermined intervals of hydrants, and one or more sprinkler units connected to the hose or hydrants. Fig. 6.2 depicts a typical sprinkler irrigation system's schematic arrangement. Fig. 6.3 shows the various sprinkler irrigation types and sprinkler head components.

6.2.3 Micro-irrigation systems

Distribution systems for micro-irrigation systems can be situated on the soil surface, beneath the surface, or hung above the ground and distribute water at low pressures. The water is applied at low rates, over long periods of time, at regular intervals, near or within the plant root zone, and at low pressure in this system. Dripper, drip, subsurface, and bubbler microirrigation systems are all examples of microirrigation [22]. Drip irrigation system consists of laterals containing emitters. Trickle irrigation is another term for drip irrigation applied to the soil surface (Fig. 6.4). A low-volume irrigation system known as surface drip irrigation or "trickle irrigation" drips water onto plants' roots at the soil's surface in small amounts. Using a field-installed system of pipes, valves, and emitters, it delivers water and other inputs (nutrients, fertilizer) gradually, regularly, and often to the roots of plants. Orchards and high value row crops like cotton, maize, sugarcane, vegetables, etc., are ideally suited for drip irrigation technique. Most people value this innovation since it increases irrigation efficiency by up to 95% while using less water and fertilizer. A standard drip system consists of the following components: a pumping unit, a fertilizer tank,

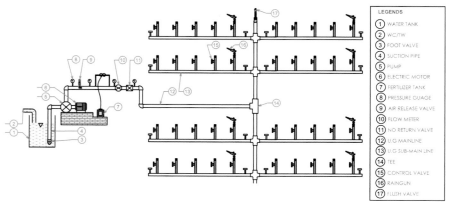

FIGURE 6.2

Schematic layout of typical sprinkler irrigation system.

FIGURE 6.3

Various sprinkler systems (A) central pivot system (B) linear move system (C) raingun system (D) parts of sprinkler head.

FIGURE 6.4

Drip irrigation installed in the field (A) orchard (B) row crop (cotton).

connecting/jointing fittings, filters, a main underground pipeline with field hydrants, header pipes, laterals, emitters, etc. Fig. 6.5 shows a typical drip irrigation system's schematic arrangement.

Initial cost of drip irrigation system is high for small farmers but the system is easy to be automated to minimize labor. A large number of studies across the world have reported that drip irrigation systems help make a significant saving (up to 50%) of water in wheat, rice, corn and other crops in comparison to flood irrigation [23–28].

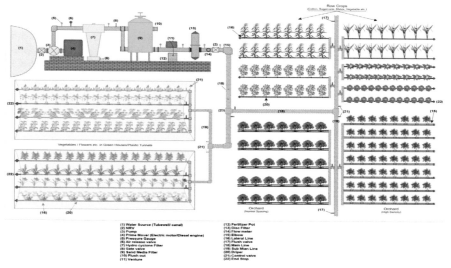

FIGURE 6.5

Schematic layout of typical drip irrigation system.

Microsystems are one of the most expensive systems of irrigation, mostly due to the expensive piping system and the severe filtration standards that are necessary. Due of the high cost and the necessity of removing the system after each growing season, row crop cultivation is generally not suitable for these systems. Subsurface drip irrigation (SDI), a technique that involves burying the laterals beneath the tillage zone, solves the second issue. Trees, shrubs, and gardens in particular benefit greatly from microirrigation. As a result of the low water application rates and minimal runoff even on steep slopes, these systems have a high efficiency and may be readily automated to save labor costs. They also provide better fertilizers and chemical applications. Initial expenditures, long-term maintenance requirements, and salt accumulation around plants are also potential drawbacks of these systems [28,29].

6.3 Literature review

6.3.1 Conceptualization of precision irrigation

Precision irrigation (PI) can be defined as the application of need-based and precisely calculated amount of water (or nutrients) to a plant (or irrigation zone) at an appropriate time and location. This irrigation system can provide the most favorable circumstances for plant/crop growth. This term (precision irrigation) is often used to describe the application of irrigation water at a variable rate; it is generally regulated by the inputs of in situ sensors. A PI system should be capable of meeting the crop's water requirement as quickly as possible, as accurately as possible, and as consistently as possible. Therefore, in order to accomplish the goal of PI, it is very important to estimate the accurate amount of crop water requirement, proper irrigation scheduling, and control of the applications to ensure that only the required amount of water/nutrients should be applied. In order to obtain the best degree of performance, a PI system should combine modern irrigation methods and application mechanisms with advanced sensing, simulations, and control technologies.

6.3.2 Benefits of precision irrigation

PI has the capability to enhance the economic efficiency by precisely adjusting irrigation inputs according to crop growth in each zone of a field and thus decreasing the production costs. Therefore, PI can reduce the water wastage while simultaneously increasing economic efficiency. Literature showed that PI can save up to 25% of water applied through conventional pressurized irrigation methods [30]. Actually, a PI system is designed to irrigate the field on the basis of its spatial and temporal variability, and thus it can save the water. A common source of variability within a field is generally caused by variations in soil characteristics and topography, which can be resulted by natural processes or stimulated by human interference (e.g., organic matter depletion, compaction, and erosion). Spatial and temporal variability of soil is commonly caused by variations in its structure; fertility; chemical, physical, and hydraulic characteristics; plant genetics; and pests and diseases. Such variations regulate soil water holding capacity, leaching of nutrients, soil and land-cover hydrologic parameters, and nutrient availability consequently influences the crop water requirement, and crop yield and quality. Precision irrigation success is dependent on providing the precise amount of water at the appropriate time and location to meet the water requirements of a crop considering the spatial and temporal variation within a field; it requires accurate understanding of the production functions of all zones within a field.

6.3.3 Main components of precision irrigation

PI is regarded as a best management strategy in precision agriculture. A PI system is composed of two major components: the physical infrastructure that provides water to crops and the control and management system that operates and manages the irrigation system. The main steps involved in the control and management system are: (1) Data collection, (2) Interpretation; (3) Control, and (4) Assessment. These steps make a cycle as shown in Fig. 6.6.

6.3.4 Data collection

A PI system requires accurate in situ spatial and temporal real-time data of soil, crop, and weather conditions. Currently, several sensors/instruments (soil and plant-based sensors and automatic weather stations) are available that can provide reliable information of required soil, crop, and weather parameters. However, the

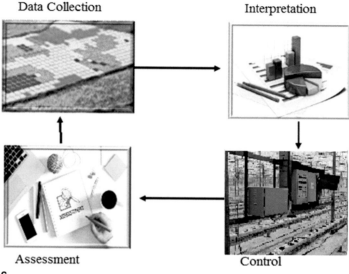

FIGURE 6.6

The cycle of precision irrigation.

installation density of such sensors/instruments in the field can influence the performance of a PI system. Some of the widely used instruments/sensors are given in Table 6.1.

6.3.5 Data interpretation/analysis

PI requires analysis and interpretation of collected data at an appropriate temporal scale and frequency. In this regard, several multidimensional modeling/simulation software/tools are available that can forecast crop response to different applications/treatments. Decision Support System for Agro-technology Transfer (DSSAT) software is widely used to simulate the crop dynamics under different climatic and irrigation conditions.

6.3.6 Control

In a precision irrigation, reallocation of inputs (water and nutrients) and establishment of the irrigation scheduling (considering the in situ soil and weather conditions) at suitable temporal and spatial scales is very important. Although amount of water applied to a field is depending on the irrigation system, there are two ways to apply need-based amounts of water to a field: by adjusting the application rate or the application time. In precision irrigation system, the most reliable and highly accurate method of managing irrigation application depths (under already established irrigation scheduling) is to integrate real-time data from on-the-go sensors with automatic controllers.

Table 6.1 Sensors and instruments that are commonly used to measure real-time soil, crop, and weather conditions.

Sr. #	Sensor name	Parameters captured
1	Hydra probe II soil sensor	Soil characteristics (temperature, moisture, salinity level, and conductivity)
2	107-L temperature sensor (BetaTherm 100K6A1B thermistor)	Plant temperature
3	237 leaf wetness sensor	Plant moisture, plant wetness, plant temperature
4	SenseH2TM hydrogen sensor	Hydrogen, plant wetness, CO_2, plant temperature
5	YSI 6025 chlorophyll sensor	Photosynthesis
6	TPS-2 portable photosynthesis	Photosynthesis, plant moisture, CO_2
7	CI-340 handheld photosynthesis	Photosynthesis, plant wetness, plant moisture, CO_2, plant temperature, hydrogen level in plant
8	CM-100 compact weather sensor	Air temperature, air humidity, wind speed, air pressure
9	Met station one (MSO)	Air humidity, air temperature, wind speed, air pressure

6.3.7 Assessment

In the PI system, evaluation, or "closing the loop," is a crucial phase. Assessing the irrigation system's agronomic, engineering, and economic performance is critical for providing feedback and optimizing the system's performance.

6.4 Existing tools and technologies for precision irrigation

PI system integrates advanced irrigation planning and application systems with sensing, simulation, and control systems. This involves the use of real-time automation and control of the system. The recent development of many advanced technologies, such as real-time positioning with GPS, proximal crop and soil sensors, variable rate technology (VRT), remote sensing, and GIS, can play a vital role in the adaptation of spatially varied irrigation or precision irrigation system. Considering this, it may create an impression that precision irrigation is all about the implementation of high-tech system for irrigation. However, this is not always true. Precision irrigation is actually a system that can be adopted to increase the crop yields by collecting and processing real-time data of the crop and the field in a systematic way.

6.5 Data acquisition tools

As already discussed, PI system requires accurate spatiotemporal information about the soil, crop, and weather conditions in the field in order to provide an optimal irrigation response. Latest PI technologies are capable to measure multiple parameters of the soil, crop, and weather (for instance, moisture content of the field, irrigation demand, and crop response) in real-time and at microscales, and can manage irrigation applications precisely. Soil moisture sensors, plant sensors, weather monitoring sensors/instruments, or combination of such sensors/instruments could be used to collect in situ data to measure spatiotemporal variability in the field conditions. Field spatial variability can be measured in three ways.

(1) Continuously (uninterrupted monitoring of field conditions with a thermal camera).
(2) Discretely (use of soil moisture probes to monitor soil moisture content).
(3) Remotely (using drones or satellite-based imageries of the field).

6.5.1 Weather data acquisition

High-quality in situ weather data is necessary for irrigation scheduling and for crop and water balance modeling. Such datasets can be obtained more affordably through networks maintained by a government organization (for instance, Pakistan Meteorological Department) or from automatic weather station (AWS) installed within the field.

6.5.2 Plant-based data sensors

Plant-based monitoring is critical for determining the impact of water shortages on crop growth and yield. A variety of plant-based sensing technologies are available to identify the onset and severity of plant stress. Plant-based sensors are classified into two types: those that require direct contact with the plant (e.g., manual or machine-mounted sensors) and those that do not require proximal contact with the plant (such as drone or satellite-mounted sensors). Manual sensors provide comprehensive temporal dynamics data of crop water status, which can help researchers/producers to understand the diurnal changes in plant health. The proximal (drone or satellite-based) sensors are better for gathering spatial data over multiple fields, farms, or regions, and hence are better for measuring geographic differences in plant stress and application of precision irrigation system.

Moisture uptake (plant water status), transpiration, and growth rate are all monitored using plant-based irrigation sensors. Crop stress is indicated by changes in these indicators, which can be considered to decide the irrigation schedule. Plant-based sensors provide no indication of how much irrigation water is required. These sensors should be used in combination with either soil moisture sensors or simulation tools to determine the irrigation demands. It is also worth mentioning that the degree of crop stress recorded is a complex mix of soil, plant, and environmental factors. As a result, the user must validate that the crop stress is due to the deficiency of soil moisture in root zone rather than the disease, pest, or unusual weather conditions.

Plant-based sensors assess the plant's water content, water potential, or physiological response to water shortages. Table 6.2 summarizes the various plant-based sensing methods.

6.5.3 Soil moisture detections

The root zone's soil moisture content varies throughout the crop field, both spatially and temporally. Because moisture is essential for plant growth and health, interpreting the variability and dynamics of soil moisture in the root zone is very important for optimal irrigation and crop management.

An electromagnetic induction sensor is a device that can be used to determine the soil's actual electrical conductivity (EC). It could be applied to monitor different variables of soil, especially soil moisture, as well as the other soil characteristics that affect EC; for instance, electrolyte concentration and texture. It has the potential to be used in irrigation management because it is easy to use, has fast speed, capability of massive data acquisition, and high volume of soil monitored when compared to other soil moisture sensors.

Table 6.2 An overview of plant-based monitoring techniques.

Plant-based measurement	Description
Plant water potential measurement sensor	
Leaf turgor pressure sensors	Can make real-time observations and analyze the water dynamics of leaves
Plant water content sensors/methods	
Stem diameter changes	Automated and sensitive to water shortages
Leaf thickness sensors	Cheap and easy to automate
Pressure bomb	Plant stress indicator that can be used in conjunction with soil moisture monitoring during different growth stages of crops
Plant water shortage response	
Thermal (proximal) sensor	A straightforward method with good spatial and temporal resolution
Xylem cavitations	Low-cost, water-stress-sensitive, and easy-to-use equipment

6.5.4 Remotely sensed data for irrigation scheduling

The use of remote sensing to determine crop characteristics for irrigation scheduling is becoming increasingly popular in irrigation. In this context, the data is frequently analyzed using a Normalized Difference Vegetation Index (NDVI) to identify changes in crop health. Alternative to the NDVI is the estimation of actual crop evaporation from remote sensing of the energy balance. Both methods provide a way to collect huge amounts of low-cost, site-specific crop evaporation information that can be used for irrigation management. The energy balance method is still in its early stages of development, whereas NDVI-based systems are in use in many regions of the world.

6.6 Challenges and opportunities

Review of the available literature on the precision irrigation has identified four crucial challenges and corresponding opportunities for improving precision irrigation system: (a) data accessibility and scalability; (b) measurement of plant water stress level; (c) simulation uncertainties and limitations; and (d) farmers involvement and motivation to use precision irrigation.

One key challenge in terms of data requirements for precision irrigation is the shortage of in-situ and proximal sensing data, both in terms of resolution and accuracy. Three challenges in data availability and scalability were identified for in-situ soil and plant data, satellite-based evapotranspiration (ET), and satellite-based vegetation and soil moisture data. On the whole, conventional in-situ sensors are either very costly or at the very least not sufficiently affordable to enable widespread adoption. In addition, they are typically required to be installed and removed before and after the growing season for row crops, resulting in an increase in labor costs. Despite the fact that in-situ sensors typically provide high-quality measurements, these measurements are taken from a single point, and as a result, they have limitations in terms of capturing spatial heterogeneity across an entire field.

More low-cost sensing tools/instruments (such as soil moisture and crop monitoring tools) must be developed to implement the precision irrigation systems in Pakistan. Literature review revealed that only few methods and technologies that use measurement on plants or plant parts to schedule irrigation are available. To define crucial values and times of plant water stress, more testing of stem water potential, IR thermography, and mobile sensing technology are needed, as well as standard measurements. To determine irrigation requirements that interface well with irrigation systems, new mobile technology is required. There is a need for further knowledge on aerodynamic and canopy resistance, as well as crop variables. There is no low-cost percolation measuring device available. To measure this loss, simple and low-cost approaches and/or models were required.

6.7 Conclusions

Water scarcity has become an urgent global issue in recent years, and it has been intensified by the growing global population and increasing dry spells in agricultural regions across the world. The inefficient use of water by irrigated agriculture has made it difficult for other industries to access this limited supply. As a result, improving irrigation water-use efficiency is critical to the sustainable food production. Smart irrigation systems have the potential to enhance irrigation efficiency, especially with the introduction of mobile networks, monitoring devices, and advanced management algorithms for precise irrigation scheduling. This chapter attempts to comprehensively synthesize the knowledge that is currently accessible regarding recent developments in precision irrigation systems using WSNs and IoT techniques. It was revealed that combining soil, plant, and weather-based monitoring systems in an irrigation scheme can significantly enhance the water-use efficiency. This information will help academicians, policy makers, and farmers to identify the optimum irrigation monitoring and management approach to enhance the irrigation scheduling in dry land agricultural regions.

References

[1] Lloret J, Sendra S, Garcia L, Jimenez JM. A wireless sensor network deployment for soil moisture monitoring in precision agriculture. Sensors 2021;21:7243. https://doi.org/10.3390/s21217243.
[2] Talaviya T, Shah D, Patel N, Yagnik H, Shah M. Implementation of artificial intelligence in agriculture for optimisation of irrigation and application of pesticides and herbicides. Artif Intell Agric 2020;4:58−73. https://doi.org/10.1016/j.aiia.2020.04.002.
[3] Kumar G. Research paper on water irrigation by using wireless sensor network. Int J Sci Eng Technol IEERT Conf Pap 2014:123−5.
[4] Shekhar Y, Dagur E, Mishra S, Tom RJ, Veeramanikandan M, Sankaranarayanan S. Intelligent IoT based automated irrigation system. Int J Appl Eng Res 2017;12(18):7306−20.
[5] Yong W, Shuaishuai L, Li L, Minzan L, Arvanitis KG, Georgieva C, et al. Smart sensors from ground to cloud and web intelligence. IFAC-PapOnLine 2018;51(17):31−8.
[6] Vij A, Vijendra S, Jain A, Bajaj S, Bassi A, Sharma A. IoT and machine learning approaches for automation of farm irrigation system. Procedia Comput Sci 2020;167:1250−7.
[7] Sidhu HS, Jat ML, Singh Y, Sidhu RK, Gupta N, Singh P, et al. Sub-surface drip fertigation with conservation agriculture in a rice-wheat system: a breakthrough for addressing water and nitrogen use efficiency. Agric Water Manag 2019;216:73−283.
[8] Savitha M, UmaMaheshwari OP. Smart crop field irrigation in IOT architecture using sensors. Int J Adv Res Comput Sci 2018;9(1):302−6.
[9] Deekshithulu NVM, Babu GR, Babu RG, Ramakrishna MS. Development of software for the microcontroller based automated drip irrigation system using soil moisture sensor. Int J Curr Microbiol App Sci 2018;7:1385−93.
[10] Al-Ghobari HM, Mohammad FS, El-Marazky MSA, Dewidar AZ. Automated irrigation systems for wheat and tomato crops in arid regions. WaterSA 2017;43:354−64.

[11] Li X, Hess TJ, Valacich JS. Why do we trust new technology? A study of initial trust formation with organizational information systems. J Strat Inf Syst 2008;17:39–71.

[12] Chandio AA, Yuansheng J, Magsi H. Agricultural sub-sectors performance: an analysis of sector-wise share in agriculture GDP of Pakistan. Int J Econ Finance 2016;8(2): 156–62. https://doi.org/10.5539/ijef.v8n2p156.

[13] Aslam M. Agricultural productivity current scenario, constraints and future prospects in Pakistan. Sarhad J Agric 2016;32(4):289–303.

[14] Bhattacharya A. Water-use efficiency under changing climatic conditions. In: Bhattacharya A, editor. Changing climate and resource use efficiency in plants. Academic Press; 2019, ISBN 9780128162095. p. 111–80. https://doi.org/10.1016/B978-0-12-816209-5.00003-9. 2019.

[15] Janjua S, Hassan I, Muhammad S, Ahmed S, Ahmed A. Water management in Pakistan's Indus Basin: challenges and opportunities. Water Pol 2021;3(6):1329–43. https://doi.org/10.2166/wp.2021.068.

[16] Kijne JW, Barker R, Molden D. Water productivity in agriculture: limits and opportunities for improvement. Wallingford, UK: CAB International; 2003.

[17] Rijsberman FR. Water scarcity: fact or fiction? Agric Water Manag 2006;80:5–22.

[18] Watto MA, Mugera AW. Groundwater depletion in the Indus Plains of Pakistan: imperatives, repercussions and management issues. Int J River Basin Manag 2016; 14(4):447–58.

[19] Bakhsh DA, Hussain Z, Sultan SJ, Tariq I. Integrated water resource management in Pakistan. In: International conference on water resources engineering & management (ICWREM-March 7–8, 2011). Lahore, Pakistan: UET; 2011. p. 7–8.

[20] FAO. Aquastat Main database. Rome, Italy: United Nations FAO; 2011.

[21] USDA. Irrigation and water management survey. Vol. 3. Special studies, Part 1. AC-17-SS-1. Washington, DC: USDA; 2019.

[22] ASABE Standards. Asae EP405.1 APR1988 (R2019): design and installation of microirrigation systems. St. Joseph, MI: ASABE; 2019.

[23] Leib BG, Hattendorf M, Elliott T, Mattews G. Adoption and adaptation of scientific irrigation scheduling: trend from Washington, USA as of 1998. Agric Water Manag 2002; 55:105–20.

[24] Wang N, Zhang N, Wang M. Wireless sensors in agriculture and food industry-recent development and future perspective. Comput Electron Agric 2006;50:1–14.

[25] Chen R, Cheng W, Liao J, Fan H, Zheng Z, Ma F. Lateral spacing in drip irrigated wheat: the effects on soil moisture, yield and water use efficiency. Field Crop Res 2015;179:52–62.

[26] Sharda R, Mahajan G, Siag M, Singh A, Chauhan BS. Performance of drip irrigated dry-seeded rice (Oryza sativa L.) in South Asia. Paddy Water Environ 2017;15:93–100.

[27] Parthasarathi T, Vanitha K, Mohandass S, Vered E. Evaluation of drip irrigation system for water productivity and yield of rice. Agron J 2018;110:1–12.

[28] Sidhu RK, Kumar R, Rana PS, Jat ML. Automation in drip irrigation for enhancing water use efficiency in cereal systems of South Asia: status and prospects. Adv Agron 2021:247–300. https://doi.org/10.1016/bs.agron.2021.01.002.

[29] Qi W, Zhang Z, Wang C, Huang M. Prediction of infiltration behaviors and evaluation of irrigation efficiency in clay loam soil under Moistube irrigation. Agric Water Manag 2021;248(106756).

[30] Hedley CB, Yule IJ. A method for spatial prediction of daily soil water status for precise irrigation scheduling. Agric Water Manag 2009;96(12):1737–45.

CHAPTER 7

Variable rate technologies: development, adaptation, and opportunities in agriculture

Shoaib Rashid Saleem[1], Qamar U. Zaman[2], Arnold W. Schumann[3], Syed Muhammad Zaigham Abbas Naqvi[4,5]

[1]*School of Engineering, University of Guelph, Guelph, ON, Canada;* [2]*Department of Engineering, Faculty of Agriculture, Dalhousie University, Truro, NS, Canada;* [3]*Citrus Research and Education Center, Institute of Food and Agricultural Sciences, University of Florida, Gainesville, FL, United States;* [4]*College of Mechanical and Electrical Engineering, Henan Agricultural University, Zhengzhou, Henan, China;* [5]*Henan International Joint Laboratory of Laser Technology in Agriculture Sciences, Zhengzhou, Henan, China*

7.1 Introduction

The world agriculture system is facing sustainability challenges due to extensive farming, deteriorating climate, water pollution, increased pest resistance to chemicals, and non-profitable agronomic practices [1]. Precision agriculture (PA) evolved over the years to replace uniform agricultural practices [2] by site-specific scientific management of farms based on the naturally existing phenomenon with the simultaneous aid of information using farm mechanization technologies in the fields. Precision farming provides a promising system that enables sustainable yield improvements, economic viability, dynamic production systems, sustainable conservation of natural resources, and limits the over-application of crop production inputs [3]. With the application of variable rate technologies (VRTs), the cost-effective deliverance of inputs without affecting crop yields has become a new trend in precision farm management. VRTs are the efficient way of precise utilization of agriculture technologies to apply the right rate of water, chemicals, nutrients, and microbes in the right place and at the right time [4].

The variable rate (VR) applicators precisely apply the seeds, fertilizers, and pesticides according to the conditions of soils, nutrient requirements, and pest attacks [5]. This management system enables the best use of the resources of advanced farm mechanization by managing fertilization, crop canopy, pest control, irrigation, soil management, and yield [6]. VRTs use two different methods for site-specific management in which one method involves the collection of data from the field first and analysis, then the recommended operations are run for VRTs in the field [7]. The other method is the on-site application of VR technologies by analyzing the field conditions in real time with the help of online sensors installed or applied within the field [7]. VR applications are comprised of different crop management actions and in each crop, from 5 up to 20 applications are identifiable such as VR planting, VR seeding, weed control, VR fertilizers, VR herbicides, and irrigation [8]. VRTs have been adopted for many cropping systems around the globe in the last 3 decades such as potato, citrus, wheat, maize, rice, and other specialty crops [9—13].

Variable rate applications are biophysically and economically effective yet their practical applications in a cropping system have a lot of prerequisite data collections for effective decision making [14]. VR applications have been adapted by many cropping systems such as potatoes to make the cost-benefit assessment and input optimization according to the site-specific needs of the agricultural food chain in the Netherlands [14]. The development of real-time sensing technologies has helped in assessing soil variability for site-specific nutrient management by replacing old and laborious manual soil sampling collection and analysis techniques [15—17]. Site-specific and subfield scale applications of VRTs play a significant role in the placement of seeds in a precise manner for cereal and specialty crops [12]. VRTs in maize have directly influenced the potential yield and environmental performance. Seed density and nitrogen applications are the determining factors of yield, nutrients leaching and evaporation in the environment, and cost efficiency in a maize cropping system [18]. Decision support systems for irrigated agriculture have been adopted extensively with high-efficiency irrigation systems and enhanced yield [19].

Variable rate irrigation control using remote sensing and spatial variability in an expert system can solve problems in irrigation [20—22]. An intelligent fuzzy inference system based on remote sensing images with speed control maps has an efficient way to control the rotation and speed of the central pivot system in a high-precision irrigation system [20]. For many fields, to assess the nitrogen needs and improved management of fertilizers application, the use of active light crop canopy reflectance sensors can override the conventional methods of single-rate fertilizers applications [21]. Biotic and biotic stress monitoring of plants for timely mitigation before the arrival of threshold levels is helped by using infrared thermocouple sensors [22]. To monitor the plant health and quantification of symptoms of infections in citrus trees, the airborne multispectral technique in the combination of VRT was developed [5,22]. Environmental impacts for nonpoint source pollution can be assessed using the PRZM-3 environmental model for the variable pesticide applications [23]. Drones and copter systems can build field maps and suggest areas with pest attacks and sense nutrient deficiencies [24]. It helps to take quick decisions and implement them in the larger field even in the capacity of smaller scales up to the single plant. Drones-based granular fertilizers spreaders with the flight controller and spread controller are improving work efficiency, reducing the production cost, minimizing the number of fertilizers used, and reducing environmental pollution. The developed control system of UAV precipitation maps can produce the highest efficiency of 0.72 ha/h and an error of less than 6.07% [25]. These VRTs help to automate and simplify the collection and analysis of information due to their user-friendly experience.

Artificial intelligence (AI) powered agriculture has made it easy to take quick measures in the field by acquiring, interpreting, and reacting to different situations [26,27]. AI-based internet of things (IoT) devices can sense, recognize, and create yield smart solutions for crop yield enhancement [26]. Robotic solutions like Rowbot have developed the best fertilizer solutions for maize crops to maximize the yield and optimized nutrient input. Further, AI-trained machines can recommend cognitive solutions for crop health monitoring, automation of irrigation applications, the readiness of fruits and crops, and sustainability [26,28]. The interface of AI for variable rate applications combines the geographic information system (GIS) and global positioning system (GPS) to incorporate fertilizers, seeds, biocides, drip lines, or sprinkler irrigation systems for their precise applications. The precise applications of seeds, fertilizers, irrigations, and biocides are based on the domain of soil, plants, and climate [27].

7.2 Variable rate technologies development

The developments in agricultural technologies in the last few decades are rising; agricultural improvements are becoming efficient. The final goal to achieve food security and better yield of crop plants is becoming a reality due to the precise management practices and field operations from tillage to harvesting [29]. The economic scenarios are also becoming cost-efficient and more profitable due to the reduced and site-specific input of the resources. Environmental stability and sustainability of the fertile land, biodiversity, and ecosystem protection targets can be achieved using the precise application of farming technologies [30].

VR methods have become more sophisticated and advanced with the development of scientific approaches to meet the needs of crops and optimize input resources. Variable rate application methods have facilitated to achieve the high quality and abundant supply of agricultural products by replacing the conventional methods, which were quite inefficient due to the continuous and same discharge rates of the input resources. The agricultural machinery is equipped with sensory systems to facilitate crop protection, ecological and environmental improvements [31]. Sensory systems with high-performance equipment including ultrasonic, infrared, and optical sensors are used in measuring, controlling, regulating, scanning, and providing information about plant growth stages and their requirements. Moreover, the sensors are very effective in monitoring and controlling the weeds by variable rate applications of biocides. Nitrogen sensors use the light reflection from plants to analyze the fertilizer requirements, that is, the more the reflection less will be chlorophyll contents in leaves, resulting in more applications of fertilizers being recommended [32].

7.2.1 Cropping systems

Management actions largely influence sustainable and profitable cropping systems. The management strategies are enabling the best use of the resources for fertilization, crop canopy, pest control, irrigation, soil management, and yield [33]. Different countries of different regions have adopted VR techniques for their specific benefits [34]. A survey report was made of input cost reduction, soil improvements, and reduction in greenhouse gas emissions, respectively [6]. Mitigation of environmental risks and increase in profitability can be achieved by the adoption of variable rate fertilizer spreaders for spot application of fertilizer in blueberry [35]. The sensing and control system containing the 6 μEye color camera was able to differentiate among the foliage, bare spots, and weeds cropping patterns in real-time. The variable rate granular was found highly significant and effective as the accuracy and the response time taken for foliage fertilizers application was just 2.25 s [35].

Subsurface water contamination studies in blueberry cropping systems revealed the input use efficiency, profitability, and reduced contamination with variable rate fertilization applications. Three variable rate applications in contrast to uniform applications of nitrogen, phosphorus, and potassium fertilizers enhanced the yield of the blueberry plants [10]. The blueberry cropping system was studied as a new method of VR application to avoid environmental contaminations and fertilizer losses through the volatilization of nitrogen fertilizer. Split variable application analysis of fertilizers in blueberry can increase the nitrogen use efficiency and significantly avoid unnecessary ammonia volatilization as compared to uniform applications [36]. The analysis of potential VR applications for potato crop management in the Netherlands identified 13 applications [14]. The applications comprised a broad spectrum including soil tillage, planting, herbicides, weed control, late blight control, nitrogen applications, selective harvest, etc. Cost efficiency and environment conservation were also enhanced by 25% due to the reduced pesticide and nitrogen usage in potato crops [17].

Combining the different precision technologies with variable rate applications can be used to predict the yield of crop plants as well. The RGB imagery data can analyze the effect of variable rate nitrogen applications. In a maize cropping system, the drone-based imaging system helped to evaluate the variable rate of nitrogen applications and successfully estimated the yield by analyzing the continuous spatial variations across the field [21]. Proximal vigor sensors can guide nutrient distribution with the help of automatic guidance machinery. A sustainable practice at a farm scale level to reduce the top dressing of nitrogen fertilizer in maize crops with variable rates was found economically profitable [37]. Compared to fixed-rate applications, the real-time VR top dress fertilizer applications through the proximal sensor did not affect or reduce the yield; moreover, they enhanced environmental stability by reducing nitrogen leaching and volatilization due to the site-specific release of fertilizers [37]. The same approach of using a proximal sensor with the combination of remote sensing has been applied in the United States in irrigated maize to manage the nitrogen and water applications at variable rates. Although nitrogen use efficiency and productivity did not show some significant results yet the nominal change in the yield was observed with the variable rate technology [38].

Predicting the needs of nitrogen fertilizers within the field in real time can be achieved using the ground-based proximal sensor with a combination of algorithmic models. An algorithmic model based on the normalized difference between vegetation index and responsive index predicted the yield and responded significantly positively to nitrogen applications at variable rates in maize crops [39]. Weather and soil conditions differentially responded to the model yet the results from normal precipitation years were more significant as compared to the years with severe drought conditions [39]. VR techniques are much effective that they can produce economic profitability under any circumstances. A case study was conducted on Australian farms to evaluate the adaptability of VR fertilizers for economic benefits in grain farms [40]. The investment on different farms varied per cropped hectare ranging from 11 to 30 dollars/ha which recovered its cost within 2–5 years with the profit rate of 28–57 dollars/ha. These profits were achieved during the years of below average crop yield [40]. The adoption rate of VR technologies in the Ontario cropping system was found significantly high in a survey. VR technology for fertilizer, seeds applications, pesticides, and fungicide precipitation were adopted by 27%, 13%, 32%, and 19%, respectively [41].

7.2.2 Sensing technologies

A sustainable environment and an economically efficient system are the major targets to achieve in the agricultural sector. The advent of modern techniques and consistent use of various sensors to apply variable inputs in production technology have made it easy to achieve precision agriculture [12]. The use of ultrasonic sensors in permanent crops, LIDAR (light detection and ranging) in tree characterization and detection, infrared sensors for characterization and detection in plants, and nitrogen sensors for nitrogen detection in plants have significantly improved the performance and quality of agricultural machinery [42]. The use of sensors in variable rate technologies enables the real-time applications of input resources according to the varying field conditions rather than the prerequisite of data collections and mappings. Data collections and mappings of field crops help to predict the crop yield, crop protection methodologies, soil variabilities and characteristics, size, age, volume, and shape in the case of tree plantations. Although the sensory system has limitations with its efficiency and accuracy in different environments and field conditions, yet different combinations of sensors and farm machinery can mitigate the accuracy of measurements [43].

The ultrasonic sensory system is used to determine the distance by using the time difference of ultrasonic waves reverting from the objects. This sensory system was initially used for industrial purposes but has been modified to use for agricultural practices on the sprayers with the combination of electromagnetic valves and control units [13,32,36] (Fig. 7.1). This system helps to work under severe conditions by minimizing the effects of humidity, dirt, fog, and high temperatures. The applications of the ultrasonic system with the combination of intermittent dispersion technique in peach and apple crops saved the input of pesticides by 10%–17% and 20%–27%, respectively. By further managing the sensors with algorithmic models, the savings of pesticides can be achieved up to 28%–34% and 36%–52% in peach and apple plants, respectively [44]. The target to reduce drift can

FIGURE 7.1

Ultrasonic sensor system [32].

also be achieved with the usage of ultrasonic sensors by determining the plant geometry and recommending the precise application of spray rates. The savings of input recourses up to 20% per nozzle can be achieved by developing the automatic spraying equipment by combining the RGB camera with an ultrasonic sensory system [45].

The highly flexible and functional sensors in all severe conditions compared to ultrasonic sensors are in use for variable rate applications and are known as optical type sensors LIDAR. Under strong magnetic fields, high temperature, electric noise, and chemical corrosion the LIDAR sensors are viable for all applications without direct contact with the objects [46]. LIDAR is a remote laser range sensor that can report the information from the larger areas with less divergent and thinner measurement beams (Fig. 7.2). In addition, LIDAR sensors can measure the 3D (dimensional) structures of plants in real-time [48]. Compared to ultrasonic sensors LIDAR is more accurate to determine the surface area, leaf mass volume, and quantification of spatial variations with the wide range of crop plants' geometric structures [48]. The data acquired for canopy information can determine the adjustment rates of agrochemical dosages, flow rates to the variations in crop canopy, and the estimation of yield in citrus plants. LIDAR sensors can produce highly accurate and detailed information about plants with high spatial resolution and detection speed and can determine the fluid loss in the form of air drift. The fitness of data obtained for plant volume and shape from LIDAR was found to be 97% accurate using 3D structural characteristics of plants as compared to the manual collections [49].

Another type of active sensor used for variable rate applications is the infrared sensor. These sensors type recognize the plant's top and obtain information on shape and density. The infrared sensor system has its best utilization in the automatic target detecting systems. Attached with the sprayers these sensors can monitor and control the airflow by increasing and decreasing the nozzle's openings and closings automatically. This way the saving of input resources up to 40% and 60%−70% vineyard drift reduction can be achieved [50]. The infrared sensors with the combination of an electrostatic spraying system can save pesticide applications ranging from 50% to

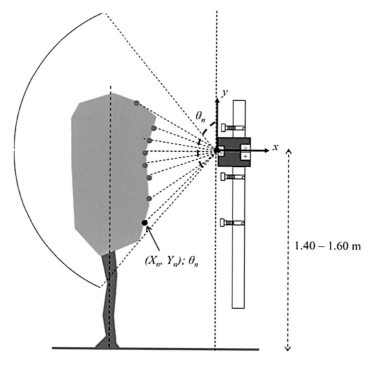

FIGURE 7.2

Process of LIDAR sensors acquiring the data from crop plants [47].

70% and utilization rate over 55%. A spot agricultural sprayer designed using the spectra-radiometry technique combined with a solenoid-controlled spray nozzle gave better results for plants with high leaf reflectivity. An infrared photoelectric switch applied with a sprayer reduced the spraying cost, increased the sensibility, anti-interference for environmental conditions, and target identification for spray was between 0.2 and 15 m [51].

Currently, the most advanced approach for variable rate spraying is the use of computer stereo vision based detection methods [52,53]. This system extracts the 3D information from digital images and the machine vision algorithms detect the gaps between plants to avoid pesticide spraying during foliage spray. This variable application technique not only has an accuracy of 90% for plant detection but also reduced pesticide usage by about 30% [52]. Machine vision smart sprayer is capable of targeted application of agrochemicals. In the blueberry cropping system, the digital color cameras coupled with boom sprayer nozzles continuously take pictures and processes in 0.15 s for spot applications of agrochemicals (Fig. 7.3). The smart sprayer not only saved the herbicide use up to 75% but also decreased premature leaves drop and enhanced the blueberry's stem girth, number of branches, and tree height [53].

FIGURE 7.3

Smart sprayer showing camera field of view for image acquisition [53].

7.3 Adaption of precision agriculture techniques

PA is encompassing five major sections based on the adoption model by the farmers for different cropping systems [54–56]. GPS, GIS, VR application, sensing technologies, and newly embraced techniques of artificial intelligence models in precision farming. The adoption of VRTs is more concentrated in developed countries such as the USA, Canada, Brazil, China, Australia, and a few other countries [54]. GPS and GIS have high rates of adoption in North America, especially for boundary mapping and sampling location mapping. Auto-steering and guidance systems are also adopted in the USA, Canada, China, and Australia in various cropping systems especially grain farming with adoption rates up to 80%. VR applicators also have high rates of adoption in the above-mentioned countries; however, the VRT's global adoption rates are still very low. The main reasons for adoption in these countries are high purchasing power, more stable crop insurance systems, and large farm sizes [54,55]. Keskin [56] conducted a review study for 47 research studies on VRTs adoption in developed countries and concluded that VRTs have huge potential in both developed and developing countries and the cost of equipment and farm sizes are the main limiting factors in the adoption of these technologies. Smart water applications in agriculture decrease irrigation water use by 8% with energy cost reductions of $7 to $13 per acre for an average US farm and increase crop production by 1.75% [57,58]. Smart irrigation management can also lower the negative impact of water stress by up to 69% in soybean cropping systems [59]. Adoption of various VRTs in corn farms in five different states in the USA showed that yield monitoring and auto steering system have the highest rate of adoption compared to soil test, which has the lowest adoption (Table 7.1) [58]. These trends should be changed

Table 7.1 Adoption of different VRTs in corn in 2010 [58].

Precision echnologies	SD	ND	NE	MN	IA
Precision agriculture used	74.5	80.2	76.3	63.5	81.7
Yield monitor used	63.2	71.1	66.9	57.1	73.4
Yield map created	37.8	34.7	36.6	39.4	46.4
Soil properties map based on: soil test	7.1	–	4.8	9.1	17.3
GPS device used to create soil properties map	23.2	14.3	15.7	24.9	33.3
VRT used for any purpose	19.4	17.9	22.8	24.4	19.7
VRT used for any fertilizing	15.3	7.9	17.3	17.4	18.6
Guidance or auto steering system used	47.6	73.4	41.3	44.8	37.2

Source: ARMS Farm Financial and Crop Production Practices/Tailored Reports: Crop Production Practices (2018).

as soil testing using proximal sensing can significantly reduce nutrient usage while optimizing crop growth.

Unmanned Aerial Vehicles (UAVs) have vast applications in agriculture from soil treatment, nutrient and pesticides/weedicides application, and physiological control and observation [60,61]. The main applications of agricultural drones are reconnaissance surveys to access soil and plant conditions, and identification of insects and weeds. The worldwide agriculture drone market value has increased dramatically from 1 million USD, and it is expected to reach 3.7 million USD by 2027 according to a survey conducted in 2019 [62].

7.3.1 User-friendliness of VRTs

Engineers and scientists have modified the existing equipment and developed new technologies over the past couple of decades. One of the main constraints in the adoption of any new technology is the user-friendly nature of that technology. It is a major constraint in terms of the agriculture industry since most of the farmers globally are less educated and require a very simple interface to operate these smart technologies. The International Organization for Standardization (ISO) defines user-friendliness as follows: "the extent to which a product can be used by specified users to achieve specified goals with effectiveness, efficiency, and satisfaction in a specified context of use" [63]. In the early 2000s, the computer and sensing systems were less user-friendly, and very few VRTs were based on plug-and-play systems [64]. The introduction of smartphones and Graphical Processing Units (GPUs) has revolutionized the adoption of these technologies [65] as farmers are now keener to use smart apps that are easy to operate. Farmers' age also plays a significant role in the adoption of modern technologies as the likelihood of accepting and operating VRTs decreases as age increases [65]. Computer knowledge limitation is another factor in the adoption of these technologies. Although newly developed apps have easy-to-use interfaces, the basic knowledge of running computer programs is necessary for sustainable agriculture [66]. Similar conclusions were also reached by researchers in Poland [67], where 100 farm owners were surveyed. They found that PA practices are more popular among farmers who are less than 40 years old, have a higher level of education, and manage larger farms study of PA technologies adoption in a survey performed on dairy farms in the USA [68]. The owners indicated the cost:benefit ratio, total investments, technical assistance, and user-friendliness as the limiting factors in the adoption of PA technologies with user-friendliness as the most important factor. Gil Moya, Koutsouris [69] researched on VR spraying equipment and suggested that 94% of farmers' highest preference is the user-friendliness of spraying equipment.

7.3.2 Artificial intelligence and variable rate technologies

Machine learning (ML) and artificial intelligence (AI) have very wide applications in the agriculture industry ranging from soil variability to yield monitoring and from disease detection to weather predictions. The main aim of using ML and AI algorithms is to attain high crop yields using minimum and optimum use of inputs such as fertilizers, pesticides, insecticides, and fungicides. Site-specific management of different cropping systems depends on better crop yield prediction. The main ML algorithms used for crop yield prediction are regression, decision trees, deep learning (DL), clustering, and artificial neural networks (ANNs) [70–72]. Researchers have used AI-ML algorithms for better prediction of soil properties, the LS-SVM least squares support vector machine (LS-SVM) method [70] and the self-adaptive evolutionary (SAE) [73], soil classification using fuzzy-logic-based Soil Risk Characterization Decision Support System (SRC-DSS) [74]; soil temperature using extreme learning machine (ELM) [29]; resolve crop selection problems using novel techniques Crop Selection Method (CSM) [75], CALEX [75], crop physical, chemical, and biological growth parameters estimation and influence on productivity using Bayesian network [76]. ANN [77,78]; disease prediction and management: computer vision system (CVS) Ji to detect multiple diseases at high speed, fuzzy logic-based database [76], an accuracy of 90% from rule-base disease detection using ANN-GIS [78], weed detection (ANN, decision tree, deep learning, and instance-based learning) [79]; weed control using invasive weed optimization (IWO) [80], big data based ANN-GA [73], support vector machine [81], etc.; effective irrigation system (when, where, and how much to irrigate) using coil rainfall, temperature, evaporation datasets based on ML adequate algorithms [82]; and harvesting (deep neural networks, data mining techniques such as k mean clustering, k nearest neighbor, ANN, and SVM) [83].

Afzaal et al. [84] used several combinations of datasets with three different convolutional neural networks (CNNs) for the classification of disease and healthy plant stages to assess and compare the classification accuracy of 2-Class, 4-Class, and 6-Class CNNs to detect early blight disease in the potato production system (Fig. 7.4). Their results suggested that machine vision and deep learning (DL)–based sprayers only in diseased areas of the potato fields have a significant potential to lower the agrochemicals in agricultural fields.

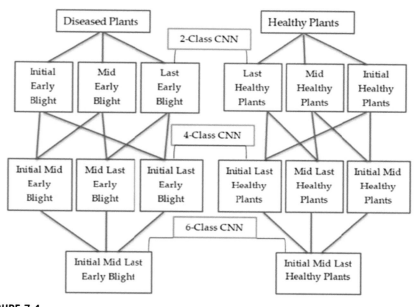

FIGURE 7.4

Concept flowchart of combinations of disease and healthy plant stages to assess and compare the classification accuracy of 2-class, 4-class, and 6-class convolutional neural networks (CNNs) [84].

A two-phase universal ML model was tested and evaluated in wheat cropping within the agroecological zones to predict crop yield [85] (Fig. 7.5).

The model was based on the development of online sequential extreme learning machines coupled with ant colony optimization (ACO-OSELM), the ACO-OSELM model was used to predict future wheat yield at six test stations [85]. The ACO-OSELM model was compared with two different models ACO-extreme learning machine (ACO-ELM) and ACO-random forest (ACO-RF), and results showed that the performance of the ACO-OSELM model was better when compared with ACO-ELM and ACO-RF models. Wheat yield can be predicted using the hybrid ACO-OSELM model in established specified crop growing areas [85].

7.4 Challenges for farmers and researchers

PA adoption at a large scale in the agriculture sector especially in developed countries is evident through the introduction of VRTs, AI-based algorithms, and UAVs. However, there are still many challenges for both researchers and farmers. These challenges include the cost of VRTs, farm size, technological limitations, lack of training, and shortage of technical services. Tey and Brindal [86] grouped approximately 34 identified factors that could influence PA adoption into seven categories

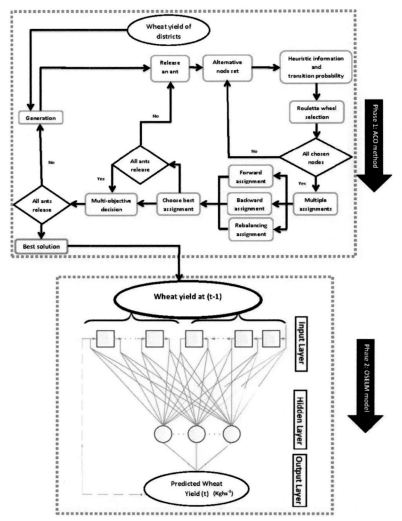

FIGURE 7.5

Flow chart of the proposed hybrid two-phase ant colony optimization algorithm integrated with the Online Sequential Extreme Learning Machine (OSELM) model [85].

namely, farmers' perceptions; institutional, informational, socioeconomic, agroecological, behavioral, and technological factors. The leading obstacle in the adaptation of PA is accurately managing daily farming operations and ensuring to adopt VRTs in all farming operations [87].

Farmers' primary focus in agriculture is farm profitability and adoption decisions of VRTs are also based on farm economics [86,88]. Additional financial resources are required to purchase these VRTS and farmers with additional nonfarm income are likely to purchase more of VRTs. However, profit uncertainty combined with high initial costs and concerns regarding feasibility can reduce the adoption desire in farmers [86]. Young farmers are likely to adopt VRTs in their farming operations compared to old farmers who are sticking with traditional ways of farming [88].

Farmers connected with producers' associations are always more educated regarding new technologies, and they gain more experience from other progressive farmers. It can not only benefit the farmers to learn about the new technologies but also reduce the negative perception of these technologies. Langyintuo, Lowenberg-DeBoer [89] reported that 63% of the farmers got positive returns and only 11% had negative returns in their review of 108 studies that should reduce the concerns of negative perceptions in the adoption of VRTs. The PA adoption trends showed a gradual increase in positive use and feedback of these VRTs over time across the globe. Extension services need to step up especially in developing countries to educate the farmers and play their role in negative perceptions of new innovative technologies.

Computers with Graphical Processing Units (GPUs) have reduced processing times significantly and made it easier for researchers to develop real-time AI-based algorithms for a wide range of applications using VRTs. These computer algorithms have their limitations and accuracies, and are based on special applications [90]. Researchers working in the agriculture sector are still facing a lot of challenges to adopt these algorithms in disease and weed predictions.

7.5 Future of variable rate technologies

VRTs introduction in agriculture through robotics, smart implements, autoguidance, and mechatronics in agriculture has reduced laborious work and minimized farm inputs using highly autonomous and smart machinery. Farm machinery has seen a great revolution from bullocks to tractors and after that the introduction of intelligent vehicles, robots, and decision support systems has revolutionized the crop production industry in the last few decades. Intelligence machines often referred to as VRTs now are being implemented in the agriculture sector to perform all types of field operations such as soil preparation, seeding, nutrient management, irrigation, weed and insect control, and crop harvesting. These VRTs have reduced the time of operation along with the cost of production and farm inputs. VRTs based on cloud computer—aided software such as GIS are being used to produce soil and crop yield maps and manage nutrients in the field. GPS is heavily used in the agriculture industry for autosteering systems in developed countries. The other application of GPS and GIS technologies in the agriculture industry is variable rate fertilization through VR applicators. These tools are being introduced in the developing world to manage crop production and achieve national and global crop production goals. However, the adoption and mass use of precision technologies in the crops depend on many factors including affordability, compatibility, and complexity of VRTs. Hands-on training to use these VRTs can reduce the farmers' perception of a "difficult to use mindset" and move toward acceptance of VRTs as the future of agriculture. The initial cost must be reduced and mass production of VRTs along with technological advances will help in the way forward in the next couple of decades.

Small land holdings are the biggest challenge in the developing world, and scientists need to come up with solutions to encourage mass adoption of these technologies. Adoption through the service providers' model is one of the solutions that can be adopted. Since these VRTs are expensive, service providers can purchase them and provide services to local farmers at cheaper rates.

Drone technology has also revolutionized the agriculture sector. Drones are not only used for reconnaissance purposes but also are widely being adopted for weed and insect management. Drones are used in standing crops in which manual or tractor-powered spraying is difficult at later growth stages. However, drones have received a mixed reception from farmers and regulators in various countries. Proper drone regulations for a specific country can increase its use and application in the agriculture sector and government and policy makers should focus on making farmer-friendly policies. The drone industry's rapid growth in the next couple of decades will encourage competition and will result in a reduction in drone prices.

Artificial intelligence is the future of PA and a combination of AI and VRTs is already making advances in crop management. Powerful computers such as GPUs have reduced the AI model training times, and the introduction of microprocessors has made it possible for engineers and technologists to adopt these intelligent algorithms on VR applicators. AI application is limited to the collection of field images and training of models. The adoption of very detailed modeling of these weeds/insects can reduce the time to take decisions but also increase the detection accuracy. Future studies should be focused on AI applications to manage not only the crop but single plant needs.

Computer models, both numerical and empirical, are being implemented in the crop and agriculture sector to predict and manage weeds/insects based on their seasonal patterns. These models can be combined with another field model to develop a decision support system. These DSS can be so powerful that farmers can increase the yield through proper and timely management of resources. DSS will play a critical role to mitigate the impacts of climate change in the coming decades. Timely dissemination of information to farmers based on intelligent weather systems and crop modeling tools can help in choosing fertilizer application, irrigation, and other farming operations using VRT applicators.

7.6 Conclusions

Traditional agriculture is barely meeting the global food demand and due to the threats of climate change, farmers across the globe need to adopt the modern agriculture technologies. Precision agriculture techniques can ensure sustainable yield, minimize over or under-application of crop production inputs, reduce cost of production, and minimize detrimental environmental impacts. Many VRTs have been developed in the last couple of decades to apply agrochemicals according to crop needs. However, the large-scale adoption remains number one challenge as farmers need constant training and willingness to adopt these technologies.

Advancements in computing capacities and introduction of GPUs have reduced the processing times of various algorithms significantly to ensure real-time applications of VRTs. The main focus of researchers in the past few years is on development of user-friendly interfaces and multi-crop programs for farmers and growers. High computing power machines are also being implemented to develop decision support systems that can predict and generate early warnings for diseases/pest attacks. Combinations of sophisticated and user-friendly VRTs along with timely dissemination of information to farmers based on intelligent weather systems and crop modeling tools can help in choosing fertilizer application, irrigation, and other farming operations in agriculture fields for sustainable agriculture production.

References

[1] Prandecki K, Wrzaszcz W, Zieliński M. Environmental and climate challenges to agriculture in Poland in the context of objectives adopted in the European green deal strategy. Sustainability 2021;13(18):10318.

[2] Roy T, George KJ. Precision farming: a step towards sustainable, climate-smart agriculture. In: Global climate change: resilient and smart agriculture. Springer; 2020. p. 199–220.

[3] West GH, Kovacs K. Addressing groundwater declines with precision agriculture: an economic comparison of monitoring methods for variable-rate irrigation. Water 2017; 9(1):28.

[4] Guerrero A, De Neve S, Mouazen AM. Chapter One: Current sensor technologies for in situ and on-line measurement of soil nitrogen for variable rate fertilization: a review. Adv Agron 2021;168:1–38.

[5] Pandey H, Singh D, Das R, Pandey D. Precision farming and its application. In: Smart agriculture automation using advanced technologies. Springer; 2021. p. 17–33.

[6] Saiz-Rubio V, Rovira-Más F. From smart farming towards agriculture 5.0: a review on crop data management. Agronomy 2020;10(2):207.

[7] Escolà i Agustí A, Arnó Satorra J, Martínez Casasnovas JA. Operation in the field: site-specific management using variable rate technologies. New Ag Int 2018;71:28–35.

[8] McNunn G, Heaton E, Archontoulis S, Licht M, VanLoocke A. Using a crop modeling framework for precision cost-benefit analysis of variable seeding and nitrogen application rates. Front Sustain Food Syst 2019;3:108.

[9] Kolady DE, Van der Sluis E, Uddin MM, Deutz AP. Determinants of adoption and adoption intensity of precision agriculture technologies: evidence from South Dakota. Precis Agric 2021;22(3):689–710.

[10] Rashid Saleem S, Uz Zaman Q, Walter Schumann A, Madani A, Ahsan Farooque A, Charles Percival D. Impact of variable rate fertilization on subsurface water contamination in wild blueberry cropping system. Appl Eng Agric 2013;29(2):225–32.

[11] Cheema MJM, Mahmood HS, Latif MA, Nasir AK. Precision agriculture and ICT: future farming. In: Developing sustainable agriculture in Pakistan. CRC Press; 2018. p. 125–36.

[12] Finger R, Swinton SM, El Benni N, Walter A. Precision farming at the nexus of agricultural production and the environment. Ann Rev Resourc Econ 2019;11(1):313−35.

[13] Zaman QU, Esau TJ, Schumann AW, Percival DC, Chang YK, Read SM, et al. Development of prototype automated variable rate sprayer for real-time spot-application of agrochemicals in wild blueberry fields. Comput Electron Agric 2011;76(2):175−82.

[14] Kempenaar C, Been T, Booij J, van Evert F, Michielsen J-M, Kocks C. Advances in variable rate technology application in potato in The Netherlands. Potato Res 2017;60(3):295−305.

[15] Burton L, Jayachandran K, Bhansali S. Review—the "real-time" revolution for in situ soil nutrient sensing. J Electrochem Soc 2020;167(3):037569.

[16] Fletcher RS, Fisher DK. Spatial analysis of soybean plant height and plant canopy temperature measured with on-the-go tractor mounted sensors. Agric Sci 2019;10(11):1486−96.

[17] Sanches GM, Magalhães PSG, Kolln OT, Otto R, Rodrigues F, Cardoso TF, et al. Agronomic, economic, and environmental assessment of site-specific fertilizer management of Brazilian sugarcane fields. Geoderma Reg 2021;24:e00360.

[18] Morris TF, Murrell TS, Beegle DB, Camberato JJ, Ferguson RB, Grove J, et al. Strengths and limitations of nitrogen rate recommendations for corn and opportunities for improvement. Agron J 2018;110(1):1−37.

[19] Ara I, Turner L, Harrison MT, Monjardino M, deVoil P, Rodriguez D. Application, adoption and opportunities for improving decision support systems in irrigated agriculture: a review. Agric Water Manag 2021;257:107161.

[20] Mendes WR, Araújo FMU, Dutta R, Heeren DM. Fuzzy control system for variable rate irrigation using remote sensing. Expert Syst Appl 2019;124:13−24.

[21] Zhang M, Zhou J, Sudduth KA, Kitchen NR. Estimation of maize yield and effects of variable-rate nitrogen application using UAV-based RGB imagery. Biosyst Eng 2020;189:24−35.

[22] Evans RG, LaRue J, Stone KC, King BA. Adoption of site-specific variable rate sprinkler irrigation systems. Irrigat Sci 2013;31(4):871−87.

[23] Du Q, Chang N-B, Yang C, Srilakshmi KR. Combination of multispectral remote sensing, variable rate technology and environmental modeling for citrus pest management. J Environ Manag 2008;86(1):14−26.

[24] Dharmaraj V, Vijayanand C. Artificial intelligence (AI) in agriculture. Int J Curr Microbiol Appl Sci 2018;7(12):2122−8.

[25] Song C, Zhou Z, Zang Y, Zhao L, Yang W, Luo X, et al. Variable-rate control system for UAV-based granular fertilizer spreader. Comput Electron Agric 2021;180:105832.

[26] Lo Piano S. Ethical principles in machine learning and artificial intelligence: cases from the field and possible ways forward. Humanit Soc Sci Commun 2020;7(1):9.

[27] Firmansyah E, Pardamean B, Ginting C, Mawandha HG, Putra DP, Suparyanto T, editors. Development of artificial intelligence for variable rate application based oil palm fertilization recommendation system. International Conference on Information Management and Technology (ICIMTech); 2021. 19−20 Aug. 2021.

[28] Naqvi SMZA, Awais M, Khan FS, Afzal U, Naz N, Khan MI. Unmanned air vehicle based high resolution imagery for chlorophyll estimation using spectrally modified vegetation indices in vertical hierarchy of citrus grove. Remote Sens Appl: Soc Environ 2021;23:100596.

[29] Fabregas R, Kremer M, Schilbach F. Realizing the potential of digital development: the case of agricultural advice. Science 2019;366(6471):3038.

[30] Movilla-Pateiro L, Mahou-Lago XM, Doval MI, Simal-Gandara J. Toward a sustainable metric and indicators for the goal of sustainability in agricultural and food production. Crit Rev Food Sci Nutr 2021;61(7):1108−29.

[31] Rockwell A D, Ayers P D. Variable rate sprayer development and evaluation. Appl Eng Agric 1994;10(3):327−33.

[32] Petrović D, Jurišić M, Tadić V, Plaščak I, Barač Ž. Different sensor systems for the application of variable rate technology in permanent crops. Tehnički glasnik 2018;12(3):188−95.

[33] Deng J, Harrison MT, Liu K, Ye J, Xiong X, Fahad S, et al. Integrated crop management practices improve grain yield and resource use efficiency of super hybrid rice. Front Plant Sci 2022;13:1−12.

[34] Castrignanò A, Buttafuoco G, Khosla R, Mouazen A, Moshou D, Naud O. Agricultural internet of things and decision support for precision smart farming. Academic Press; 2020.

[35] Chattha HS, Zaman QU, Chang YK, Read S, Schumann AW, Brewster GR, et al. Variable rate spreader for real-time spot-application of granular fertilizer in wild blueberry. Comput Electron Agric 2014;100:70−8.

[36] Abbas A, Uz Zaman Q, Walter Schumann A, Brewster G, Donald R, Shafqat Chattha H. Effect of split variable rate fertilizationon ammonia volatilization in wild blueberry cropping system. Appl Eng Agric 2014;30(4):619−27.

[37] Schillaci C, Tadiello T, Acutis M, Perego A. Reducing topdressing N fertilization with variable rates does not reduce maize yield. Sustainability 2021;13(14):8059.

[38] Dahal S, Phillippi E, Longchamps L, Khosla R, Andales A. Variable rate nitrogen and water management for irrigated maize in the Western US. Agronomy 2020;10(10):1533.

[39] Tagarakis AC, Ketterings QM. Proximal sensor-based algorithm for variable rate nitrogen application in maize in northeast U.S.A. Comput Electron Agric 2018;145:373−8.

[40] Robertson M, Carberry P, Brennan L. The economic benefits of precision agriculture: case studies from Australian grain farms. Crop Pasture Sci 2007;60:2012.

[41] Mitchell S, Weersink A, Erickson B. Adoption of precision agriculture technologies in Ontario crop production. Can J Plant Sci 2018;98(6):1384−8.

[42] Kerry R, Escolà A, Mulla D, Gregorio Lopez E, Llorens Calveras J, Lopez A, et al. Sensing approaches for precision agriculture. Springer; 2021.

[43] Delavarpour N, Koparan C, Nowatzki J, Bajwa S, Sun X. A technical study on UAV characteristics for precision agriculture applications and associated practical challenges. Rem Sens 2021;13(6):1204.

[44] Giles DK, Delwiche MJ, Dodd RB. Sprayer control by sensing orchard crop characteristics: orchard architecture and spray liquid savings. J Agric Eng Res 1989;43:271−89.

[45] Hočevar M, Širok B, Jejčič V, Godeša T, Lešnika M, Stajnko D. Design and testing of an automated system for targeted spraying in orchards. J Plant Dis Prot 2010;117(2):71−9.

[46] Zhang B, Xie Y, Zhou J, Wang K, Zhang Z. State-of-the-art robotic grippers, grasping and control strategies, as well as their applications in agricultural robots: a review. Comput Electron Agric 2020;177:105694.

[47] Llorens J, Gil E, Llop J, Escolà A. Ultrasonic and LIDAR sensors for electronic canopy characterization in vineyards: advances to improve pesticide application methods. Sensors 2011;11(2):2177−94.

[48] Llorens J, Landers A. Variable rate spraying: digital canopy measurement for air and liquid electronic control. Aspect Appl Biol 2014;(122):1−8.

[49] Guo Q, Su Y, Hu T, Guan H, Jin S, Zhang J, et al. Lidar boosts 3D ecological observations and modelings: a review and perspective. IEEE Geosci Remote Sens Mag 2021; 9(1):232−57.

[50] Gregorio E, Torrent X, Planas de Martí S, Solanelles F, Sanz R, Rocadenbosch F, et al. Measurement of spray drift with a specifically designed lidar system. Sensors 2016; 16(4):499.

[51] Jiao J-S, Chu JY, Ma GB. Application of infrared photoelectrics switch for spraying on aspen. J Agric Mech Res 2005;3:216−7.

[52] Luo L, Tang Y, Zou X, Wang C, Zhang P, Feng W. Robust grape cluster detection in a vineyard by combining the adaboost framework and multiple color components. Sensors 2016;16(12):2098.

[53] Esau T, Zaman Q, Groulx D, Farooque A, Schumann A, Chang Y. Machine vision smart sprayer for spot-application of agrochemical in wild blueberry fields. Precis Agric 2018; 19(4):770−88.

[54] Derpsch R, Friedrich T, Kassam A, Li H. Current status of adoption of no-till farming in the world and some of its main benefits. Int J Agric Biol Eng 2010;3(1):1−25.

[55] Mariano MJ, Villano R, Fleming E. Factors influencing farmers' adoption of modern rice technologies and good management practices in the Philippines. Agric Syst 2012;110:41−53.

[56] Say SM, Keskin M, Sehri M, Sekerli YE. Adoption of precision agriculture technologies in developed and developing countries. Online J Sci Technol 2018;8(1):7−15.

[57] Gralla P. Precision agriculture yields higher profits, lower risks. Hewlett Packard Enterp 2018.

[58] Uddin MM. Factors influencing adoption and adoption intensity of precision agriculture technologies in South Dakota. South Dakota State University; 2020.

[59] Paz JO. Analysis of spatial yield variability and economics of prescriptions for precision agriculture: a crop modeling approach. Iowa State University; 2000.

[60] Krishna KR. Push button agriculture: robotics, drones, satellite-guided soil and crop management. CRC Press; 2017.

[61] Huuskonen J, Oksanen T. Soil sampling with drones and augmented reality in precision agriculture. Comput Electron Agric 2018;154:25−35.

[62] Ben Ayed R, Hanana M. Artificial intelligence to improve the food and agriculture sector. J Food Qual 2021;2021:5584754.

[63] Giua C, Materia VC, Camanzi L. Management information system adoption at the farm level: evidence from the literature. Br Food J 2021;123(3):884−909.

[64] Brisco B, Brown RJ, Hirose T, McNairn H, Staenz K. Precision agriculture and the role of remote sensing: a review. Can J Rem Sens 1998;24(3):315−27.

[65] Shadrin D, Menshchikov A, Somov A, Bornemann G, Hauslage J, Fedorov M. Enabling precision agriculture through embedded sensing with artificial intelligence. IEEE Trans Instrum Meas 2020;69(7):4103−13.

[66] Daberkow SG, McBride WD. Farm and operator characteristics affecting the awareness and adoption of precision agriculture technologies in the US. Precis Agric 2003;4(2): 163−77.

[67] Biczkowski M, Jezierska-Thöle A, Rudnicki R. The impact of RDP measures on the diversification of agriculture and rural development—seeking additional livelihoods: the case of Poland. Agriculture 2021;11(3):253.

[68] Borchers MR, Bewley JM. Producer assessment of precision dairy farming technology use, pre-purchase considerations, and usefulness. J Dairy Sci Submitted 2014;98(6).

[69] Gil Moya E, Koutsouris A, Balsari P, Codis S, Nuyttens D, Fountas S, editors. Exploring the adoption of innovative spraying equipment. Portugal: Farming Systems Facing Climate Change and Resource Challenges; 2020.

[70] Ben Ayed R, Ennouri K, Ben Amar F, Moreau F, Triki MA, Rebai A. Bayesian and phylogenic approaches for studying relationships among table olive cultivars. Biochem Genet 2017;55(4):300—13.

[71] Elavarasan D, Vincent DR, Sharma V, Zomaya AY, Srinivasan K. Forecasting yield by integrating agrarian factors and machine learning models: a survey. Comput Electron Agric 2018;155:257—82.

[72] Zhang C, Liu J, Shang J, Cai H. Capability of crop water content for revealing variability of winter wheat grain yield and soil moisture under limited irrigation. Sci Total Environ 2018;631—632:677—87.

[73] Nahvi B, Habibi J, Mohammadi K, Shamshirband S, Al Razgan OS. Using self-adaptive evolutionary algorithm to improve the performance of an extreme learning machine for estimating soil temperature. Comput Electron Agric 2016;124:150—60.

[74] Di Vaio A, Boccia F, Landriani L, Palladino R. Artificial intelligence in the agri-food system: rethinking sustainable business models in the COVID-19 scenario. Sustainability 2020;12(12):4851.

[75] Li M, Yost RS. Management-oriented modeling: optimizing nitrogen management with artificial intelligence. Agric Syst 2000;65(1):1—27.

[76] Ji B, Sun Y, Yang S, Wan J. Artificial neural networks for rice yield prediction in mountainous regions. J Agric Sci 2007;145(3):249—61.

[77] Lal H, Jones JW, Peart RM, Shoup WD. Farmsys—a whole-farm machinery management decision support system. Agric Syst 1992;38(3):257—73.

[78] Crop nutrition diagnosis expert system based on artificial neural networks. In: Haiyan S, Yong H, editors. Third international conference on information technology and applications (ICITA'05); 2005. 4—7 July 2005.

[79] Liakos KG, Busato P, Moshou D, Pearson S, Bochtis D. Machine learning in agriculture: a review. Sensors 2018;18(8).

[80] Morellos A, Pantazi X-E, Moshou D, Alexandridis T, Whetton R, Tziotzios G, et al. Machine learning based prediction of soil total nitrogen, organic carbon and moisture content by using VIS-NIR spectroscopy. Biosyst Eng 2016;152:104—16.

[81] Crop selection method to maximize crop yield rate using machine learning technique. In: Kumar R, Singh MP, Kumar P, Singh JP, editors. International conference on smart technologies and management for computing, communication, controls, energy and materials (ICSTM); 2015. 6—8 May 2015.

[82] Goap A, Sharma D, Shukla AK, Rama Krishna C. An IoT based smart irrigation management system using Machine learning and open source technologies. Comput Electron Agric 2018;155:41—9.

[83] Sadgrove EJ, Falzon G, Miron D, Lamb DW. Real-time object detection in agricultural/remote environments using the multiple-expert colour feature extreme learning machine (MEC-ELM). Comput Ind 2018;98:183—91.

[84] Afzaal H, Farooque AA, Schumann AW, Hussain N, McKenzie-Gopsill A, Esau T, et al. Detection of a potato disease (early blight) using artificial intelligence. Rem Sens 2021;13(3):411.

[85] Ali M, Deo RC, Xiang Y, Prasad R, Li J, Farooque A, et al. Coupled online sequential extreme learning machine model with ant colony optimization algorithm for wheat yield prediction. Sci Rep 2022;12(1):5488.

[86] Tey YS, Brindal M. Factors influencing the adoption of precision agricultural technologies: a review for policy implications. Precis Agric 2012;13(6):713–30.
[87] McBratney AB, Minasny B, Whelan BM. Obtaining 'useful' high-resolution soil data from proximally-sensed electrical conductivity/resistivity (PSEC/R) surveys. Precis Agric 2005;5:503–10.
[88] Adrian AM, Norwood SH, Mask PL. Producers' perceptions and attitudes toward precision agriculture technologies. Comput Electron Agric 2005;48(3):256–71.
[89] Langyintuo AS, Lowenberg-DeBoer J, Faye M, Lambert D, Ibro G, Moussa B, et al. Cowpea supply and demand in West and Central Africa. Field Crop Res 2003;82(2):215–31.
[90] Colyer SL, Evans M, Cosker DP, Salo AIT. A review of the evolution of vision-based motion analysis and the integration of advanced computer vision methods towards developing a markerless system. Sports Med Open 2018;4(1):24.

CHAPTER 8

Yield monitoring and mechanical harvesting of wild blueberries to improve farm profitability

Karen Esau[1], Qamar U. Zaman[1], Aitazaz A. Farooque[2,4], Travis J. Esau[1], Arnold W. Schumann[3], Farhat Abbas[4]

[1]Department of Engineering, Faculty of Agriculture, Dalhousie University, Truro, NS, Canada; [2]Faculty of Sustainable Design Engineering, University of Prince Edward Island, Charlottetown, PE, Canada; [3]Citrus Research and Education Center, Institute of Food and Agricultural Sciences, University of Florida, Gainesville, FL, United States; [4]Canadian Centre for Climate Change and Adaptation, University of Prince Edward Island, Charlottetown, PE, Canada

8.1 Introduction

Wild blueberry (*Vaccinium angustifolium* Ait.) is a small fruit crop that is managed on a biyearly cycle to maximize yields. The first year sees vigorous vegetative growth followed by a fruit production year. The fields are pruned following the harvest of the crop to maximize crop growth and productivity. About 90% of the wild blueberry fields are harvested using mechanical harvesters, while the remaining is still hand-raked using manual labor. Hand-raking is highly labor-intensive, posing a challenge to harvesting significant acreage within the short span of the harvesting season. Mechanical harvesters for wild blueberries are developed as an efficient alternative to accomplish harvesting of thousands of acreages over the short season. Mechanical harvesting has its challenges as wild blueberries are naturally grown with no specific rows or trim lines on an undulating topography, causing significant variations in fruit losses. The combination of mechanical (ground speed, head revolutions, and harvester age), climatic (temperature, humidity, and time of harvest), crop (plant height, fruit zones, and density), and field characteristics (soil moisture, slope, and elevation) play a major role in fluctuating the fruit losses and quality during the harvest. The quality of the harvested berries remains a major criterion for judging the performance of the harvesting approach. Fruit quality is an important prospect for a crop that is being sold in international markets. This chapter demonstrates the importance of precision harvesting technologies to optimize yield and minimize losses during mechanical harvesting. The wild blueberry plants loaded with fruit and the measurement of canopy wetness using a sensor before harvest is shown in Fig. 8.1.

FIGURE 8.1

Wild blueberry plants loaded with fruit (*left*) and measurement of canopy wetness using a sensor before harvest (*right*).

8.1.1 Cultivation and arvesting of wild blueberries

Wild blueberry fields are established on deforested farmland by removing competing vegetation [1]. The newly developed fields can have a significant portion of weeds/grass patches and bare spots. In some fields, the presence of weeds and disease is observed to be about 50% of the total field [2]. The canopy of wild blueberries expands by an underground rhizome system that is 70–85% of the total weight of the plant [3]. The wild blueberry plant's height usually varies from 5 to 30 cm [4]. Harvesting of the wild blueberry crop is performed when approximately 90% of the berries' color changes to blue, which is weather dependent. In eastern Canada, the harvesting of wild blueberries usually starts in early August and lasts for a month. Wild blueberries must be harvested before frost occurs in early September, as frost damages the fruit quality. Most of the wild blueberry acreage in eastern Canada is harvested mechanically. Inadequate harvesting operation by ignoring the spatial variations in soil, crop, and climate can result in poor harvesting efficiency and fruit quality (i.e., damaged, soft, and leaky berries). These berries are at increased risk of decay during postharvest storage if they are damaged, soft, and leaky. Wild blueberries need to be harvested at an appropriate ripening and maturity stage to improve berry quality. Different researchers used color changes as an indication of the fruit ripening and maturity of berry crops. Cepons and Stretch [5] observed color changes from light green to light brown and then dark brown, as cranberry fruit matured. Naczk and Shahidi [6], reported that violet, blue, and red colors in most vegetables, fruits, and cereals were due to anthocyanins pigments.

Hand-raking is a historical manual harvesting method for wild blueberries. Hand-rake resembles the design of a cranberry scoop. Hand-raking has several drawbacks when it comes to wild blueberry harvesting, that is, labor requirements, interference with weed, and experience of hand-raking. Hand-raking in weedy fields can deteriorate the quality of the harvested crop [4]. Weed presence and interference with hand-rake harvesting within wild blueberry fields affect fruit quality and increase fruit losses [4]. These challenges and constraints associated with the hand-raking emphasized the need to develop a mechanical harvester for efficient harvesting with reduced losses. The design of a mechanical harvester for the wild blueberry crop must consider several challenges including variable plant height, bare soil, rough terrain, presence of debris and weeds, the timing of harvest, fruit maturity, and short harvest season. The research work on the design of a mechanical blueberry harvester was started in the 1950s, but a practical harvester was not developed until the 1980s [7]. Many researchers attempted to develop and modify existing mechanical harvester designs in the last few decades [8–11] to come up with an effective and efficient design that can increase harvesting efficiency and reduce berry losses. The first commercially available harvester was the modified form of a cranberry picker with six raking combs that racked in the opposite direction of the machine. This design was not practical as it caused soil digging and high fruit losses during harvesting. In 1979, the harvester design was improved by Doug Bragg Enterprises (DBE) Limited, Collingwood, Nova Scotia. The DBE modified the original design by Dale et al. [12] to improve the picking efficiency during harvesting. Farooque et al. [13] evaluated the wild blueberry harvester to quantify the fruit losses. They suggested optimal mechanical parameters to increase harvesting efficiency and reduced berry losses. Ali et al. [14] investigated the impact of harvesting times (early, mid, and late seasons) on wild blueberry fruit quality. Jameel et al. [15] explored the role of plant growth parameters on the picking performance of the wild blueberry harvester. These developments and evaluations facilitated the adoption of the mechanical harvester in eastern Canada's wild blueberry industry. A comparison of hand-raking with the mechanical harvester is depicted in Fig. 8.2.

FIGURE 8.2

Hand-raking (*left*) and mechanical harvesting of wild blueberries (*right*).

8.2 Factors affecting ripening of wild blueberry

Maturity indices are used as a decision-making factor to determine when the crop should be harvested to get the maximum desirable quality for the consumers [16]. Most of the attributes (appearance, fruit quality, nutritional value, and flavor) that draw the consumers' attraction and compositional changes associated with fruit ripening can occur during maturity and fruit development. Robertson et al. [17] revealed that the harvest maturity of fruit can be judged using indices, such as fruit color, size, shape, firmness, the concentration of soluble solids, and titratable acidity. The correlation between fruit surface color and other ripening attributes suggested that these attributes were found to be the same within a cultivar, location, season, harvest dates, and berry sizes [18]. Physical appearance is not the only determinant of fruit maturity; flavor, quality, titratable acidity, and soluble solid contents have also been considered as a criterion for fruit maturity [16]. Research has shown that the total soluble solids increased but titratable acids declined in blue and ripened wild blueberries [19]. El-Nemr et al. [20] observed 85.4% moisture and a significant number of total solids, anthocyanin content, and declining sugar levels in fresh pomegranate. The key ripening characteristics of strawberries were found to be the organic acid, soluble solids, and anthocyanin concentration. The wild blueberry fruit quality is a function of several integrated factors (mechanical, crop, soil, harvest season, and weather conditions) to meet consumer needs. Efficient harvesting systems that are capable of accounting for the spatial and temporal variability in real-time have a strong potential to improve picking efficiency and fruit quality.

8.2.1 Meteorological factors

Weather conditions are an important variable when it comes to wild blueberry harvesting. Mechanical harvesting in extremely hot and humid conditions facilitates the deterioration of fruit quality as fruit firmness declines. Firmness is one of the most significant characteristics in marketing fresh blueberries and attracting consumers in international markets [11]. Variations in fruit firmness are a useful and dependable indicator of ripening behavior [21,22], as they are related to weather conditions. Mechanical harvesting in wet conditions results in the blockage of weeds in the harvester head, causing cut-split and damaged berries. The decline in fruit firmness usually occurs during fruit ripening on the plant due to fluctuations in weather conditions. The firmness also decays in processing and storage, which is a physiological phenomenon [23,24]. Significant quantities of wild blueberries are discarded as their firmness falls below the retail market norm [25]. Salvador et al. [26] reported that firmness is an essential trait for the commercialization of persimmon. Several researchers have established a relationship between blueberry firmness and moisture loss [27,28], as a function of weather conditions. This implies that climatic conditions, particularly ambient air temperature and relative humidity, have a significant impact on the loss of water from wild blueberry fruit, consequently impacting the firmness and ripening. This situation illustrates the need to perform mechanical harvesting at the right time of ripening with adequate climatic conditions to improve berry quality for regional, national, and international markets. The recording of weather parameters and crop characteristics during the mechanical harvesting of wild blueberries are illustrated in Fig. 8.3.

8.3 Precision harvesting technologies

Precision agriculture technologies provide a set of tools to monitor and map soil, crop, and fruit characteristics that may provide considerable information to automate the mechanical harvesting operation. An automated harvesting system comprising yield monitors, sensors, and control systems has the potential to account for spatial variation during harvesting to facilitate the adjustments in mechanical parameters of the harvester to increase harvesting efficiency and quality. The harvesting efficiency and fruit quality are impacted by the plant height, soil conditions, topography, clones, presence of weeds and bare spots, ground speed, head revolutions, wear and tear of teeth bars, and climatic variables [29]. Precision agriculture tools can provide an opportunity to sense these attributes during harvesting for efficient harvesting. Farooque et al. [30] integrated an ultrasonic sensor, a real-time global positioning system (RTK-GPS), a digital camera, and a slope sensor onto the mechanical harvester to estimate plant height, fruit yield, elevation, and slope in real-time during harvesting. They also examined the efficiency of the harvester with these sensed attributes. Farooque et al. [4] quantified the harvesting losses using different mechanical parameters to suggest an optimal combination of operational settings to increase berry picking efficiency. Zaman et al. [31] developed a digital photography-based system to estimate wild blueberry fruit yield using custom software that calculates the blue pixels in the geo-referenced images in real-time. The results of this suggested the use of digital color photography in a combination of load cells to map

FIGURE 8.3

Recording weather parameters (*left*) and crop characteristics (*right*) during mechanical harvesting of wild blueberries.

and monitor the blueberry fruit yield. Jameel et al. [32] examined the joint impact of plant and fruit characteristics on wild blueberry fruit losses and quality using precision agriculture technologies. Farooque et al. [13] modeled the fruit losses as a function of mechanical, soil, crop, and yield variables using artificial neural networks (ANN) to suggest the optimal scenarios for efficient harvesting within wild blueberry fields. Das [33] employed machine vision and deep learning technologies to assess the fruit quality at different conveyer belts during mechanical harvesting. Best crop management practices (selective fertilization, fungicides, herbicides, pollination, etc.) have resulted in an increased yield within wild blueberry fields [34] demanding for efficient harvesting system to be developed and implemented to increase profit margins for growers. Precision harvesting systems can be developed by integrating machine vision, 3D cameras, deep learning, artificial intelligence, machine learning, load cells, yield monitoring, height sensors, slope sensors, and optical devices into the mechanical harvesters. These technologies once integrated into wild blueberry harvesters have a strong potential to improve fruit yield and quality, thus increasing profit margins for growers.

8.3.1 Challenge with hand-raking

Over the last century, hand-raking has remained a popular harvesting method for wild blueberries. The major concern with the hand-raking is the harvesting loss that can vary from 15% to 20% within wild blueberry fields [4], depending on the weedy and weed-free fields. Fruit losses in weed-free fields were reported to be lower compared to weedy fields. The availability of trained labor is a concern when it comes to the hand-raking of wild blueberries. Non-experienced hand-rakers tend to cause more fruit losses when compared to trained ones. Commercial wild blueberry production in conjunction with a shortage of labor, short harvesting season, and fruit losses associated with hand-raking, has raised sustainability concerns with hand-raking to harvest commercial fields. This challenging situation requires the implementation and adoption of efficient mechanical harvesters for the wild blueberry industry to cover the acreage being harvested, with very little dependency on manual labor, over the short harvesting season to improve profit margins for growers.

8.3.2 History of mechanical harvesters

Mechanical harvesters are used by the wild blueberry industry over the last two decades dues to the challenges mentioned above. However, the harvesting efficiency and fruit quality seem to be a concern for the industry. The first commercial harvester was developed by altering the cranberry picker with six raking combs that racked in the opposite direction of the machine. This design was not practical as it caused high fruit losses and soil digging during the harvesting [12]. The CRCO-UM blueberry harvester was evaluated by Towson [35] who reported 85% efficiency depending on field conditions. In 1979, DBE Limited worked on the improvement of the harvester design to reduce losses and improve fruit quality. They incorporated head rotational speed hydraulic control systems for head height adjustments, speed controls of belts and conveyors, cleaning brush rotation speed, and altering the width of the picking head for efficient harvesting.

Today, over 80% of the wild blueberry acreage is harvested mechanically. The fields with undulating terrain and rocky nature are still hand-raked. Hall et al. [7] reported the efficiency of the DBE harvester to be 68% in weedy fields, and 75% in weed-free fields, which was found to be comparable with hand-raking. Sibley [36] reported that the mechanical harvester was 69% effective based on engineering evaluation. Farooque et al. [30] integrated a digital camera into the wild blueberry harvester to predict pre-harvest fruit losses by comparing the yield obtained by the cameras when compared with the actual yield from various wild blueberry fields. They highlighted the necessity to evaluate harvesters and quantify the fruit losses during harvesting. Farooque et al. [4] conducted a detailed evaluation of the DBE mechanical harvester in varying field conditions (slope, soil type, and soil moisture), crop characteristics (plant density, height, and fruit zones), mechanical parameters (different levels of ground speed and head revolutions), and climatic conditions to quantify the fruit losses and suggest an optimal combination of these variables to increase harvesting efficiency with minimal fruit losses during harvesting. These recommendations are currently being used by the wild blueberry industry to facilitate an increase in profit margins through increased harvesting efficiency.

8.3.3 Working principle of wild blueberry mechanical harvester

The mechanical harvesters are commonly operated by mounting on tractors to harvest wild blueberries. DBE limited is one the biggest suppliers of mechanical harvesters to the wild blueberry industry. These mechanical harvesters (single, double, and multi head versions) have high durability and the capacity to cover large acreage during the harvest season. The picking reel's functioning concept during mechanical harvesting is illustrated in Fig. 8.4. The operator manually operates the picking head which is driven by a hydraulic motor from the tractor cockpit. Sixteen teeth bars are attached to the head of the harvester and each bar has 67 equally spaced curved teeth. These teeth bars are operated in the opposite direction of the tractor's ground speed during mechanical harvesting. The hydraulic motor allows the operator to control the harvester head's rotation speed. The change in revolutions per minute facilitates altering the upward movement of the teeth through the plants for effective picking of wild blueberries.

The operator of the harvester needs to be proactive and alert to change the harvesting settings on the go to minimize fruit losses during mechanical harvesting. Controlling tractor ground speed, head height, and head revolutions by considering the spatial variations encountered by the harvester in real-time, is a stressful task for the operator. Additionally, the operator must keep an eye on the conveyors, cleaning brush, and empty and filled bins, which makes this operation even more stressful for the operator. The cleaning brush in the head is operated in the direction of the picking reel to get rid

130 CHAPTER 8 Yield monitoring and mechanical harvesting

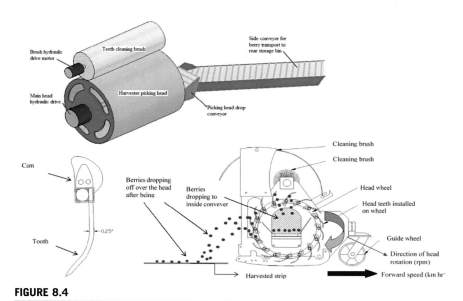

FIGURE 8.4

Schematic diagrams to show the working principle of a commercial wild blueberry harvester (Farooque et al. [4]).

of dirt and debris before the berries land on the inner conveyor. The harvested wild blueberries are delivered from the inner conveyor to the side conveyor, which deposits them in the bin behind the tractor. Before storing material in a bin, the blower fan located at the conveyor is used to remove any debris and foreign material. The operator must manually change the harvester settings to accommodate for spatial variability in plant height, topography, fruit zone, soil conditions, and plant density to efficiently harvest wild blueberries, which is a quite tiring and stressful operation for the harvester. Additionally, accommodating these changes in real time requires a skilled operator. A nonskilled operator would cause higher fruit losses and damage to fruit quality. This situation demands the development and implementation of precision harvesting technologies, comprising sensing and control systems, to be integrated into the wild blueberry harvester to accommodate for the adjustments in the harvester settings in real time to increase harvesting efficiency and fruit quality. Additionally, these on-the-go adjustments in harvesting operational parameters based on sensing and control system have a strong potential to lower the operator's stress during mechanical harvesting. Furthermore, precision harvesting technologies will eliminate the dependency on the operator's skills for the efficient harvesting of wild blueberries.

8.4 Factors affecting the mechanical harvesting

Several factors influence the mechanical harvesting of wild blueberries. These factors include crop maturity, berry ripening, time of harvest, soil moisture, leaf wetness, harvesting season (early, mid, or late), plant parameters, topography, and operator skills. The ultimate goal is to optimize the harvesting factors to attain the best fruit quality with reduced losses. The harvesting of wild blueberry at the proper time and stage is immensely valuable for adequate fruit quality. The late harvest can result in more fruit losses and fruit quality damage. Early harvesting can harvest unripened berries, resulting in poor fruit quality and marketability. The typical time of harvest starts when about 90% of blueberries turn blue (in early to mid-August) and the harvesting season remains about three to 4 weeks in North America [4]. Significant research has been conducted to quantify the berry-picking efficiency of various harvesting methods; however, the effect of meteorological factors and field conditions on the physical berry quality has not been explored. Research suggested that the scientific comparative assessment of different harvesting approaches is needed to aid farmers, processors, and stakeholders in making informed decisions when pursuing the harvesting of wild blueberries for the fresh market.

The temperature and relative humidity in the field affects the wild blueberry fruit losses and the quality of the harvested product. Storage of the harvested product by leaving the filled bins in hot ambient air temperature affects the postharvest storage quality, as berries are softened due to exposure to direct heat. These observations can be used to develop best management practices to maximize berry quality and minimize fruit losses during mechanical harvesting. The other factors causing the fluctuations in fruit losses and berry quality are canopy wetness, berry surface temperature, berry size, plant characteristics, and fruit firmness. After the harvest, the berries are characterized into four categories, that is, good berries, bruised, cut split, and mixed with debris. Ali et al. [14] evaluated the impact of four temperature ranges at the time of harvest (TH) on berry quality during harvesting. The selected ranges in their study were TH-I ($\leq 20°C$), TH-II ($20.1-25°C$), TH-III ($25.1-29.9°C$), and TH-IV ($\geq 30°C$). Statistical analysis of extensive data at a range of TH combinations identified the optimal scenarios to achieve good marketable berries to increase profit margins. This research also indicated the worst combinations with bruised, soft skin and/or mixed with foreign materials, cut-split berries; and unripened small or shrunk berries, to suggest optimal settings for the industry. A representative sample to characterize the berry quality during harvesting is presented in Fig. 8.5.

8.4.1 Weather-related factors

Weather conditions play a major role during wild blueberry harvesting. Research conducted by Ali et al. [37] has shown that irrespective of the harvesting method the relative humidity was significantly different at the time of harvest (TH-I to TH-IV). The average berry temperature before harvesting for four temperature ranges at the time of harvest was significantly lower than the respective ambient air temperatures. The temperature at the time of harvest affected the berry quality and its marketability. Different harvesting methods can have a significant effect on temperature variations on percent good and percent bruised berries, while the percent cut-split berries and percent debris remained unaffected by the temperature at harvest. These results

FIGURE 8.5

A harvested raw sample of wild blueberries (shown in a horizontally placed sample collection tray on top of the photo) sorted into (A) good berries (marketable berries without any bruise and/or foreign materials), (B) unfavorable bruised berries having soft and/or damaged skin, (C) cut-split berries (poor berries having badly ruptured skin), and (D) debris comprising foreign materials and off-color small or shrunk berries.

suggested the need to perform the mechanical harvesting at an appropriate time to ensure better fruit quality and profitability. The reaction of harvesting methods with temperature and humidity variations can be different as each method has its own mechanical and human factors. The recording of the berry surface temperature using a thermal sensor, soil moisture, and plant density measurements during mechanical harvesting of wild blueberries are presented in Fig. 8.6.

FIGURE 8.6

Recording the berry surface temperature using a thermal sensor (*left*) and soil moisture and plant density measurements (*right*) during mechanical harvesting of wild blueberries.

8.4.2 Human-induced factors—operator's skills

The harvesting of wild blueberries is highly dependent on the operator's ability and skills. A nonskilled operator can have significantly higher fruit losses when compared with trained operators. Although the mechanical harvesting of wild blueberries remains the most effective method, but the unavailability of experienced operators during harvest season emphasizes the need to automate the wild blueberry harvester using precision agriculture tools to avoid dependence on skilled operators. A fully automated harvester can perform the harvester adjustments on the go by considering the spatial variation to improve harvestable yield. Esau et al. [34] developed a control system that can offer electronic signal feedback to the harvester picking reel for automated height adjustment. They suggested that this system can be integrated into five harvester heads at the same time. In addition to the development of a control system, this research evaluated the implementation of three enhancements (a baseline function, a tandem movement function, and an one-to-one function) for an operator to operate numerous heads in real-time to obtain full automation of the wild blueberry harvester. The evaluation of the precision and accuracy of each of these functions yielded absolute mean deviations of 3.10 mm (a tandem movement function), 2.20 mm (a baseline function), and 2.50 mm (one-to-one function). Both hydraulic and electric actuators were investigated for their performance in this system, but the second (electric) actuator was found to be sluggish for commercial harvesters and therefore not practical. Under the same load, the electric actuator needed 13.96 s to reach the entire stroke (203.20 mm) required by the harvester head, whereas the hydraulic actuator needed only 2.40 s. These systems and research investigations are extremely important and provide a foundation block to develop precision harvesting technologies to optimize operator's stress, improve field and harvesting efficiency, and improve berry quality and profitability for the wild blueberry industry.

8.4.3 Field topography and vegetative conditions

Topographic factors affect the harvesting operation within wild blueberry fields. Field terrain was considered one of the important criteria in designing and developing the wild blueberry harvester. The performance of the harvester varies with the topographic features, as head position and distance from the ground surface vary. Main terrain characteristics that influence the harvesting efficiency include the existence of bare soil patches, uneven field surfaces, and the presence of rocks and stumps in the root zone. Zaman et al. [38] observed a 0.80–31.0 degree slope variation within wild blueberry fields. These variations in slope can affect the harvesting efficiency and berry quality as it would be hard to perform manual adjustments in the harvester by considering these steep changes. This extreme slope variation within fields necessitates the need for the development of an automated harvester. Variations in topography are translated into changes in plant height and the existence of the fruit zones as mechanical harvester scoops through the blueberry plants during harvesting. Farooque et al. [4] discovered the height of fruit zones from 74.0 to 346.0 mm above the ground in various wild blueberry fields, with a change in topography. To account for the variations in plant height and fruit zones, integrated with the spatial changes in slope, during mechanical harvesting requires a control system to make the adjustments in harvesting operation in real time. Additionally, these considerations should be integrated into the new design of the control system as the industry moves toward multiheaded machines for harvesting commercial acreage in eastern Canada.

8.4.4 Mechanical factors

A wide range of mechanical factors can have a significant influence on the picking performance of the harvester and fruit quality. These factors include the age of the harvester, wear and tear of teeth bars, deterioration of cleaning brush, the distance of cleaning brush from teeth bars, faulty blower fan, inadequate conveyor speeds, too-high ground speed of the harvester, and inadequate header revolutions. The mechanical harvester needs to be properly maintained to make sure all its components are working properly before starting the harvesting operations for effective operations. Farooque et al. [4] evaluated the wild blueberry harvester suggesting that an inadequate combination of ground speed and header revolutions can cause significantly higher losses during the mechanical harvesting of wild blueberries. The results of their study indicated a linear relationship between fruit losses with fruit yield, suggesting adjusting the ground speed and header revolutions based on the spatial variations in fruit yield. Effective maintenance of the harvester combined with optimal operational settings by considering spatial and temporal variations can improve berry picking efficiency by increasing the harvestable yield and quality.

8.5 Fruit losses during harvesting

The picking efficiency of blueberry harvesters has utilized a criterion to assess the picking performance of the harvester. The higher picking efficiency illustrates lower fruit losses and vice versa. Research studies conducted during a wide period

[4,35,36] described the importance of the harvester evaluation to enhance berry picking efficiency. Farooque et al. [4] studied the impact of header rotations and ground speeds on the picking efficiency of the harvester equipped with 16 harvesting teeth bars. They showed significantly lower fruit losses and improved harvesting efficiency at 26 revolutions per minute of the head speed at a ground speed of 1.2 km per hour ground speed. Visual observations revealed that the losses on the ground after harvesting were significantly higher than the losses on the shoots and those caused by the blower, which agreed with the findings of Farooque et al. [4]. Extensive evaluations by Farooque et al. [4] showed a linear trend between fruit losses and fruit yield, that is, fruit losses were found to be higher in high-yield areas and vice versa. Depending on field circumstances, the efficiency of the mechanical harvester ranges between 85% and 90% [4]. Based on the extensive literature search, it is recommended to adjust the operational setting of the harvester by considering the spatial variations to improve harvesting efficiency and lower fruit losses during mechanical harvesting.

Yield losses during mechanical harvesting are caused by multiple factors, that is, mechanical parameters, operator abilities, meteorological conditions, crop traits, field topography, and soil structure [30,39]. Optimizing the harvest operation with the above-mentioned factors is key to maximizing the fruit yield and quality [40]. The picking efficiency is affected by inherently nonlinear and complex interactions, emphasizing the need to employ modeling techniques to predict and forecast fruit losses [41]. Modeling a network of relationships is a resilient, adaptable, and scalable method that offers a selection of learning algorithms to optimize these nonlinear operations. The prediction system designed from the inherent variables is preferable, as it considers the spatial variations. In situations when inputs and outputs are inherently variable, a system designed for prediction is preferable. Hence, a system that can learn and train from repeated field trials is practical. This system will undoubtedly become more dependable over time and can adjust to unanticipated changes to optimize harvesting operations.

8.5.1 Prediction of yield losses

The ANN can be employed efficiently for nonlinear functional modeling [42], such as wild blueberry harvesting operations with significant spatial and temporal variabilities. These networks can outperform the statistical models, especially in nonlinear multiple processing systems. The ANN architecture is like the human brain's structure, which learns the link between input and output variables and builds a hidden layer for knowledge transfer in terms of a complicated set of connected weights [43]. Depending on the task at hand, the structure and topology of the ANN might be incredibly simple or complicated. The ANN is made up of several basic processing units known as nodes. These are connected by weight function and direct communication lines. The ANN is organized into three layers: the input layer (observations), the hidden layer, and the output layers (conclusions). The ANN models can solve complicated problems easily and concisely. The practicality and functionality of ANN require the users to have very little theoretical expertise. The ANN can conduct optimization, generalization, relation, approximation, prediction, abstraction, adaptation, and classification of a complex system [44], as a function of various variables. The ANN has been widely used to forecast the growth stages [45], flood forecasting [46], yield prediction and disease estimation [47], rainfall-runoff predictions [48], agrochemicals assessment [49], water level prediction [50], and streamflow estimations [51]. Farooque et al. [13] modeled the fruit losses during mechanical harvesting as a function of mechanical, crop, and yield parameters to propose optimal settings for wild blueberry harvesting.

8.5.2 Yield mapping and mitigation of fruit losses

The integration of a digital color camera into a mechanical harvester to predict fruit yield before harvesting to estimate fruit losses nondestructively is extremely valuable to identify the sources of berry losses. The georeferenced camera technology coupled with load cells can provide an accurate yield mapping system for the wild blueberry cropping system. Research trials have been conducted on nondestructive fruit yield estimates for diverse cropping systems. Schumann et al. [52] created a ground-based ultrasonic sensor system combined with digital photography to measure the features of citrus orchard trees and yield in real time. The performance of a cost-effective yield monitoring system was examined by Zaman et al. [31] for a wild blueberry cropping system. They examined the performance of a cost-effective 10 megapixel digital color camera to estimate the wild blueberry yield using custom-developed software. Results showed that image processing coupled with digital photography can be used to calculate the blue pixel ratio for fruit yield estimation. The automated yield monitoring system developed by Zaman et al. [38] consisted of an RTK-GPS, a digital color camera, software, and a ruggedized laptop computer. They evaluated these yield monitoring systems and developed georeferenced yield maps to showcase the variations in wild blueberry fruit yield. These research studies provided scientific pieces of evidence to explore the way to design and automate the wild blueberry harvester.

Wild blueberry fields have had their fruit yield increase by a factor of two to three because of improved management practices [53]. The average fruit yield between 1969 and 1974 was 960 kg/ha, as calculated by Metzger and Ismail [54]. Farooque et al. [4] examined an average yield of 8000 kg/ha in a well-managed wild blueberry field. However, from 1985 to 1989, the average production in selected wild blueberry fields was 1580 kg/ha [55]. The fruit losses with mechanical harvesting were found to be more than 10% for high-yield fields using optimal harvesting settings. There is a notion that the fruit yield is directly proportional to the harvesting losses during mechanical harvesting. Several factors can be considered to mitigate the fruit losses and improve berry picking efficiency, that is, adequate maintenance of the mechanical harvesters, accounting for the spatial and temporal variations in soil, plant and terrain characteristics, appropriate time, and climatic conditions for harvest, and adjusting the machine parameters in relations to the spatial variations.

In a technology world, it is important to develop a fully automated harvester to increase harvestable yield, improve fruit quality, and generate higher profit margins for the wild blueberry industry.

8.6 Conclusions

The wild blueberry crop is unique. The mechanical harvesting of wild blueberries comes with various challenges in terms of rough terrain, fluctuations in fruit yield, operator skills, variations in plants and soil characteristics within fields, interference of weeds and bare patches during the harvesting, harvester maintenance, time of harvest, and climatic conditions. All these challenges jointly contribute to increased fruit losses and poor berry quality during mechanical harvesting. The optimization of the harvester by considering all the factors, and adjusting its operating conditions, can facilitate the development and implementation of the best harvester settings to improve its picking performance, harvestable yield, and quality. The development of precision harvesting technologies by integrating sensing and control systems into the wild blueberry harvester can help to accommodate the adjustments in the harvester settings in real time, by considering the variabilities encountered, to increase harvesting efficiency and fruit quality. A fully automated mechanical harvester has a strong potential to lower the operator's stress during mechanical harvesting. Furthermore, precision harvesting technologies will eliminate the dependency on the operator's skills for the efficient harvesting of wild blueberries.

References

[1] Eaton LJ. Nutrient cycling in lowbush blueberries. Halifax: Dalhousie University; 1988.
[2] Zaman QU, Schumann AW, Percival DC. An automated cost-effective system for real-time slope mapping in commercial wild blueberry fields. HortTechnology April 1, 2010;20(2):431—7.
[3] Jeliazkova and D, Percival E. Effect of drought on ericoid mycorrhizae in wild blueberry (*Vaccinium angustifolium* Ait.). Can J Plant Sci 2003;83(3):583—6.
[4] Farooque AA, Zaman QU, Groulx D, Schumann AW, Yarborough DE, Nguyen-Quang T. Effect of ground speed and header revolutions on the picking efficiency of a commercial wild blueberry harvester. Appl Eng Agric 2014;30(4):535—46.
[5] Ceponis MJ, Stretch AW. Berry color, water-immersion time, rot, and physiological breakdown of cold-stored cranberry fruits. Hortscience 1983;18(4):484—5.
[6] Naczk M, Shahidi F. Phenolics in cereals, fruits and vegetables: occurrence, extraction, and analysis. J Pharmaceut Biomed Anal 2006;41(5):1523—42.
[7] Hall IV, Craig DL, Lawrence RA. A comparison of hand raking and mechanical harvesting of lowbush blueberries. Can J Plant Sci 1983;63(4):951—4.
[8] Abdalla DA. Raking and handling lowbush blueberries. Cooperative Extension Service; 1963.
[9] Grant DC, Lamson BA. Berry picking machine. In: U. S. Patent No. 367,692; 1972.

[10] Rhodes RB. The harvesting of lowbush blueberries. American Society of Agriculture Engineers; 1961. p. 61–206.
[11] NeSmith DS, Prussia S, Tetteh M, Krewer G. Firmness losses of rabbiteye blueberries (*Vaccinium ashei* Reade) during harvesting and handling. In: VII international symposium on vaccinium culture, vol. 574; December 4, 2000. p. 287–93.
[12] Dale AE, Hanson J, Yarborough DE, McNicol RJ, Stang EJ, Brenan R, et al. Mechanical harvesting of berry crops. In: Janick J, editor. Horticultural reviews, vol. 16. John Wiley & Sons; 1994. p. 255–382.
[13] Farooque AA, Zaman QU, Nguyen-Quang T, Groulx D, Schumann AW, Chang YK. Development of a predictive model for wild blueberry harvester fruit losses during harvesting using artificial neural network. Appl Eng Agric 2016;32(6):725–38.
[14] Ali S, Zaman QU, Schumann AW, Udenigwe CC, Farooque AA. Quantification of wild blueberry fruit losses at different time intervals during mechanical harvesting. In: 2015 ASABE annual international meeting 2015. American Society of Agricultural and Biological Engineers; 2015. p. 1.
[15] Jameel MW, Zaman QU, Schumann AW, Quang TN, Farooque AA, Brewster GR, et al. Effect of plant characteristics on picking efficiency of the wild blueberry harvester. Appl Eng Agric 2016;32(5):589–98.
[16] Kader AA. Fruit maturity, ripening, and quality relationships. In: International symposium effect of pre-& postharvest factors in fruit storage, vol. 485; 1997. p. 203–8.
[17] Robertson JA, Meredith FI, Lyon BG, Norton JD. Effect of cold storage on the quality characteristics of 'Au-Rubrum' plums 1. J Food Qual 1991;14(2):107–17.
[18] Kushmann LJ, Ballinger WE. Effect of season, location, cultivar, and fruit size upon quality of light-sorted blueberries1. J Am Soc Hortic Sci 1975;100(5):564–9.
[19] Forney CF, Kumudini UK, Jordan MA. Effects of postharvest storage conditions on firmness of 'Burlington' blueberry fruit. In: 8th North American research and extension workers conference, wilmington, North Carolina. Proceedings, Wilmington/NC; 1998. p. 227–32.
[20] El-Nemr SE, Ismail IA, Ragab M. Chemical composition of juice and seeds of pomegranate fruit. Food Nahrung 1990;34(7):601–6.
[21] Crisosto CH, Crisosto GM, Ritenour MA. Testing the reliability of skin color as an indicator of quality for early season 'Brooks' (*Prunus avium* L.) cherry. Postharvest Biol Technol 2002;24(2):147–54.
[22] Crisosto CH, Garner D, Crisosto GM, Bowerman E. Increasing 'Blackamber' plum (*Prunus salicina* Lindell) consumer acceptance. Postharvest Biol Technol 2004;34(3):237–44.
[23] Abbott JA. Quality measurement of fruits and vegetables. Postharvest Biol Technol 1999;15(3):207–25.
[24] Chen P. Quality evaluation of technology of agricultural products. In: Proceedings of the Korean society for agricultural machinery conference 1996. Korean Society for Agricultural Machinery; 1996. p. 171–90.
[25] Prussia SE, Tetteh MK, Verma BP, NeSmith DS. Apparent modulus of elasticity from FirmTech 2 firmness measurements of blueberries. Transac ASABE 2006;49(1):113–21.
[26] Salvador A, Arnal L, Monterde A, Carvalho CP, Martínez-Jávega JM. Effect of harvest date in chilling-injury development of persimmon fruit. In: International conference postharvest unlimited downunder 2004; 2004. p. 399–400.

[27] Miller WR, McDonald RE, Melvin CF, Munroe KA. Effect of package type and storage time-temperature on weight loss, firmness, and spoilage of rabbiteye blueberries. Hortscience 1984;19(5):638–40.

[28] Tetteh MK, Prussia SE, Nesmith DS, Verma BP, Aggarwal D. Modeling blueberry firmness and mass loss during cooling delays and storage. Transac ASAE 2004;47(4):1121.

[29] Farooque AA, Zaman QU, Schumann AW, Madani A, Percival DC. Response of wild blueberry yield to spatial variability of soil properties. Soil Sci 2012;177(1):56–68.

[30] Farooque AA, Chang YK, Zaman QU, Groulx D, Schumann AW, Esau TJ. Performance evaluation of multiple ground based sensors mounted on a commercial wild blueberry harvester to sense plant height, fruit yield and topographic features in real-time. Comput Electron Agric 2013;91:135–44.

[31] Zaman QU, Schumann AW, Percival DC, Gordon RJ. Estimation of wild blueberry fruit yield using digital color photography. Trans ASABE 2008;51(5):1539–44.

[32] Jameel MW. Effect of crop characteristics and machine parameters on berry losses during wild blueberry harvesting. Doctoral dissertation; 2015.

[33] Das AK. Development of an automated debris detection system for wild blueberry harvesters using a convolutional neural network to improve fruit quality. 2020.

[34] Esau TJ, Zaman QU, Chang YK, Schumann AW, Percival DC, Farooque AA. Spot-application of fungicide for wild blueberry using an automated prototype variable rate sprayer. Precis Agric 2014;15(2):147–61.

[35] Towson AL. CRCO-UM blueberry harvester evaluation. In: International engineering report. Niagra Falls, NY: Chisholm Ryder Co. Inc; 1969.

[36] Sibley JK. Wild blueberry harvesting technologies: engineering assessment. Can Agric Eng 1994;35:33–9.

[37] Ali S. Effect of harvesting time on berry losses during mechanical harvesting of wild blueberries (Doctoral dissertation). 2016.

[38] Zaman QU, Schumann AW, Percival DC. An automated slope measurement and mapping system. HortTechnology 2010;20(2):431–7.

[39] Bryant CR, Smit B, Brklacich M, Johnston TR, Smithers J, Chiotti Q, et al. Adaptation in Canadian agriculture to climatic variability and change. In: Societal adaptation to climate variability and change 2000. Dordrecht: Springer; 2000. p. 181–201.

[40] Fritz D, Weichmann J. Influence of the harvesting date of carrots on quality and quality preservation. Sympos Qual Veg 1979;93:91–100.

[41] Chen CR, Ramaswamy HS, Alli I. Prediction of quality changes during osmo-convective drying of blueberries using neural network models for process optimization. Dry Technol 2001;19(3–4):507–23.

[42] Park JH, Huh SH, Kim SH, Seo SJ, Park GT. Direct adaptive controller for nonaffine nonlinear systems using self-structuring neural networks. IEEE Trans Neural Network 2005;16(2):414–22.

[43] Setiono R, Leow WK, Thong J. Opening the neural network black box: an algorithm for extracting rules from function approximating artificial neural networks. ICIS Proc 2000: 17.

[44] Kung SY, Diamantaras K, Mao WD, Taur JS. Generalized perceptron networks with nonlinear discriminant functions. In: Neural networks: theory and applications; 1992. p. 245–79.

[45] Clapham WM, Fedders JM. Modeling vegetative development of berseem clover (*Trifolium alexandrinum* L) as a function of growing degree days using linear regression and neural networks. Can J Plant Sci 2004;84(2):511–7.

[46] Wright NG, Dastorani MT. Effects of river basin classification on artificial neural networks based ungauged catchment flood prediction. In: Proceedings of the international symposium on environmental hydraulics; 2001.
[47] Batchelor WD, Yang XB, Tschanz AT. Development of a neural network for soybean rust epidemics. Transac ASAE 1997;40(1):247−52.
[48] Sobri H, Amir Hashim MK, Nor Irwan AN. Rainfall-runoff modeling using artificial neural network. In: Proceeding of 2nd world engineering congress. Kuching; 2002.
[49] Yang CC, Prasher SO, Sreekanth S, Patni NK, Masse L. An artificial neural network model for simulating pesticide concentrations in soil. Transac ASAE 1997;40(5):1285−94.
[50] Patrick AR, Collins WG, Tissot PE, Drikitis A, Stearns J, Michaud PR, et al. Chapter 13.7: Use of the ncep mesoeta data. In: A water level predicting neural network; 2002.
[51] Wright NG, Dastorani MT, Goodwin P, Slaughter CW. A combination of neural networks and hydrodynamic models for river flow prediction. In: Fifth international conference on hydroinformatics; 2002.
[52] Schumann AW, Hostler K, Melgar JC, Syvertsen JP. Georeferenced ground photography of citrus orchards to estimate yield and plant stress for variable rate technology. In: Proceedings of the Florida state horticultural society. vol. 120; 2007. p. 56−63.
[53] Yarborough DE. Factors contributing to the increase in productivity in the wild blueberry industry. Small Fruits Rev 2004;3(1−2):33−43.
[54] Metzger HB, Ismail AA. Management practices and cash operating costs in lowbush blueberry production [Maine]. Bulletin-University of Maine; 1976.
[55] DeGomez T, Forsythe HY, Lambert D, Osgood E, Smagula J, Yarborough D. Introduction to growing blueberries in Maine. Wild Blueberry Fact Sheet 1990:220.

CHAPTER 9

Artificial intelligence and deep learning applications for agriculture

Travis J. Esau[1], Patrick J. Hennessy[1], Craig B. MacEachern[1], Aitazaz A. Farooque[2], Qamar U. Zaman[1], Arnold W. Schumann[3]

[1]Department of Engineering, Faculty of Agriculture, Dalhousie University, Truro, NS, Canada; [2]Faculty of Sustainable Design Engineering, University of Prince Edward Island, Charlottetown, PE, Canada; [3]Citrus Research and Education Center, Institute of Food and Agricultural Sciences, University of Florida, Gainesville, FL, United States

9.1 Artificial intelligence, machine learning, and deep neural networks

Artificial intelligence is a form of problem solving which uses computer algorithms to mimic human decision-making. The birth of this field is generally considered [1] to have occurred in 1956 at a workshop at Dartmouth College [2]. Attendees of the workshop attempted to determine how to make computers solve complex problems. They operated on the assumption that every aspect of learning and intelligence could be precisely programmed so a computer could simulate it.

Machine learning algorithms are a form of artificial intelligence that use data to learn how to make decisions. Mitchell [3] gives the following definition of machine learning:

"A computer program is said to learn from experience E with respect to some class of tasks T and performance measure P, if its performance at tasks in T, as measured by P, improves with experience E."

The tasks T, in machine learning are often a form of classification or regression. For example, an image of a plant leaf could be classified as healthy or unhealthy. The performance P, of this task is the accuracy or the error rate of classified leaves. The experience E is a dataset containing images of healthy and unhealthy plant leaves.

Machine learning algorithms are broadly categorized into supervised learning, unsupervised learning, and reinforcement learning. Supervised machine learning algorithms used labeled datasets to create predictions. In general, data x with labels y are used to teach the supervised learning algorithm such that it can predict y for new data [4]. Unsupervised machine learning algorithms find patterns within datasets, which can be used for applications such as outlier detection [5] and clustering [6]. Reinforcement learning algorithms interact with an environment and use a feedback loop for learning [7].

9.2 Machine learning approaches
9.2.1 Supervised machine learning

Linear regression is a simple form of machine learning. The task T, of linear regression is to predict y given x using $y = W \times x + b$. The experience E, are the data points in x, y coordinates. The coefficient W and bias b must be found such that y can accurately be predicted from x. The mean squared error (MSE) of the predicted and actual y values can be used as the performance P, with a goal of minimizing the MSE.

Support vector machines (SVMs) are a classification algorithm which fits a hyperplane between data classes [8]. Data with known classes are plotted and the SVM fits a hyperplane which separates the data points with one class on each side of the plane. The SVM maximizes the distance, or margin, from the hyperplane to the nearest data point in each class using "support vectors," which are parallel to the hyperplane (Fig. 9.1).

Decision trees are useful for classification or regression. The treelike architecture consists of decision and leaf nodes in a parent and child structure, originating from a root decision node [9,10]. Each parent decision node splits input data into one of its corresponding child nodes depending on whether a certain condition is met

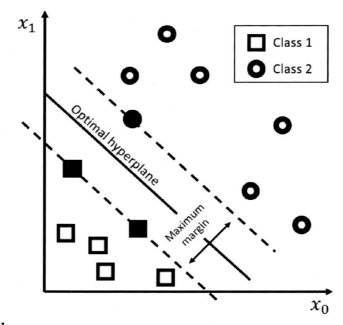

FIGURE 9.1

A support vector machine (SVM) with two independent variables, x_0 and x_1, which groups data into two classes.

(Fig. 9.2). The conditions are learned such that data points are separated into distinct classes. Each decision node, except the root decision node, is also a child of a previous decision. Every decision creates a new branch, or subtree, within the decision tree.

The k-Nearest Neighbors (kNN) algorithm is used to perform classification or regression on data by comparing it to existing data. The distances from a new data point to its k closest nearby points are measured [11,12]. This algorithm is simple to use [13], but computationally expensive [14].

Ensemble methods combine multiple machine learning models to produce a result. A popular example of this is the random forest algorithm, which uses multiple decision trees to achieve a result [15]. Random subsets of the training data are used to develop multiple decision trees, which helps prevent overfitting to training data.

Perceptrons are binary classifiers, created by Rosenblatt in 1958 [16], inspired by biological neurons, which store and process information in mammalian brains. In machine learning applications, they consist of multiple inputs x_n, scaled with weights w_n, which are multiplied together and the products are summed [17] (Fig. 9.3). If the sum of the products is greater than 0, the output y of the perceptron is 1. Otherwise, the output of the perceptron is zero. In their 1969 book *Perceptrons: an introduction to computational geometry*, Minsky and Papert [17] provide proofs of the capabilities and limitations of perceptrons. Notably, they prove that single layers of perceptrons cannot execute the "exclusive or" (XOR) logical operation, which led to a decline in artificial intelligence research until the 1980s [18]. However, multiple layers of perceptrons can be combined to achieve this function. This is the basis for artificial neural networks (ANNs), which will be discussed later in this chapter.

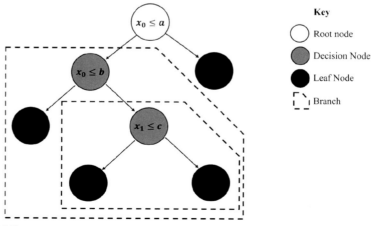

FIGURE 9.2

A decision tree which splits data based on conditions of two independent variables, x_0 and x_1.

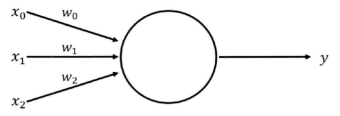

FIGURE 9.3

A perceptron with three inputs and a single output.

9.2.2 Unsupervised machine learning

K-Means clustering is a method of grouping data [6]. Data is plotted and cluster centers are randomly initialized. The distances between points and each center are calculated, then each point is assigned to the nearest center. The location of a cluster center is recalculated as the mean of all points assigned to it. This process is repeated until convergence.

Fuzzy C-Means clustering allows data points to belong to more than one cluster [19,20]. An advantage of Fuzzy C-Means is that it can produce more accurate results using less data than the other unsupervised learning models. Because of this, Rehman et al. [21] suggest that the use of Fuzzy-C Means is preferable for data that is ambiguous or between the center of clusters. The main disadvantage it has to the other models is that it is computationally heavy, so it requires more computational time to produce accurate results. It also requires a limited amount of noise and interference. Gaussian Mixture Models (GMMs) are a "soft" clustering algorithm, similar to Fuzzy C-Means. They group data points into clusters using Gaussian probability distributions [22].

9.3 Deep neural network approaches
9.3.1 Artificial neural networks

As discussed in Section 9.2.1, the perceptron is an artificial model for storage and processing of information in mammalian brains [16]. Artificial neural networks (ANNs) are a popular machine learning technique created from many perceptrons. The perceptrons (often called "nodes" or "neurons") are organized into layers of inputs, hidden calculations, and outputs. The outputs from one layer of neurons are used as inputs for the next layer of neurons. ANNs with more than one hidden layer are considered deep neural networks.

Rumelhart et al. [23] developed backpropagation of errors to train ANNs. This method involves feeding preclassified data through the ANN, then adjusting the weights and biases between the neurons proportionally to the degree of error in the prediction by the ANN. Optimization algorithms based on gradient descent [24] are used to minimize the error of the ANN during backpropagation training.

Graphics processing units (GPUs) are typically used to train ANNs because of their ability to perform many calculations in parallel [25].

9.3.2 Convolutional Neural Networks

Convolutional Neural Networks (CNNs) are deep ANNs which transform data with linear and nonlinear transformations. A hidden convolutional layer, consisting of three sublayers, is used to automatically extract features and patterns from images [4,26]. The components of a convolutional layer are.

1. Convolution operation
2. Nonlinear activation function
3. Pooling function

The first sublayer, the convolution operation, transforms data linearly using a kernel or filter [4]. The filter is an array, typically of size 2×2 to 4×4, which modifies groups of neurons simultaneously. The values stored in the filter are determined through backpropagation training. The second sublayer transforms the data using an activation function. Often, a Rectified Linear Unit (ReLU) is used to raise the value of all negative numbers to zero, leaving positive numbers unchanged [27,28]. Alternatives to the ReLU include the sigmoid, the leaky ReLU, and the Exponential Linear Unit (ELU). The third sublayer, the pooling function, reports a summary of the values stored in a group of nearby neurons, which helps to smooth noise from the inputs [4,29]. Most commonly, a max pooling function [30], which outputs the maximum value of all relevant neurons, is used.

LeCun et al. [31] developed a CNN to recognize handwritten digits in images originally employed for computers to read bank cheques. The method became popular for general image analysis after Krizhevsky et al. [32] used a CNN to win the 2012 ImageNet Large Scale Visual Recognition Challenge (ILSVRC) [33]. Since 2015, CNNs have been increasingly used in agriculture for image recognition tasks including weed detection, fruit counting, classifying fruits based on ripeness and deterioration, and crop classification [34,35].

9.4 Machine learning applications in agriculture

A review in 2018 by Liakos et al. [14] found 40 peer-reviewed journal articles published since 2004 on machine learning applications in agriculture. They divided the applications into four main categories: crop management, livestock management, soil management, and water management. The majority of articles published at that time related to crop management, with livestock management being the second most common. Because of this, they broke these two categories down further into subcategories that highlight more specific uses of machine learning applications in agriculture. Crop management was subdivided into yield prediction, disease detection, weed detection, crop recognition, and crop quality.

Livestock management was subdivided into animal welfare and livestock production.

A subsequent review in 2021 by Benos et al. [35] used the same categorization technique as Liakos et al. [14]. Benos et al. [35] found 338 peer-reviewed articles relating to applications of machine learning in agriculture. The percentage of articles relating to soil and water management did not change between 2018 and 2021, but the share of research on crop management increased at the expense of livestock management (Table 9.1).

9.4.1 Crop management

In 2017, Ramos et al. [36] conducted research on the implementation of machine learning for crop yields. Their method used a machine vision system that could predict the yield, maturation percentage, and weight of coffee fruits on a branch by analyzing images of them. The purpose of using the machine learning algorithm is to better inform the farmer of when the crop is harvestable and which sections of a field may need more resources to get the crop to a harvestable state.

Machine learning algorithms that use reinforced learning models can be used for automation in the agricultural field. Sharma et al. [37] noted that it can be in the form of "robot navigation, machine skill acquisition, and real-time decision-making." A review by Rehman et al. [21] expands on this by stating that agricultural robots can be a solution to the downward trend of farm workers out in the field. By opening the door to agricultural robots, landowners can use them for crop management to increase crop yield and decrease the spread of crop diseases.

In 2019, Rehman et al. [32] stated that the Discriminant Analysis (DA) model has been used in the past for detecting weeds. The benefit of farmers using machine

Table 9.1 Share of machine learning in agriculture research in 2021 and 2018.

Application	Percentage of articles in 2021 [35] (%)	Percentage of articles in 2018 [14] (%)
Crop management	68	61
Yield prediction	20	20
Disease detection	19	22
Weed detection	13	8
Crop recognition	13	3
Crop quality	3	8
Livestock management	12	19
Animal welfare	7	7
Livestock production	5	12
Soil management	10	10
Water management	10	10

learning algorithms for weed and disease detection is that they can improve their strategy of where they must apply agrochemicals in the field to avoid an abundance of weeds and spread of crop diseases. Through applying agrochemicals only where they are needed, farmers can benefit both economically and environmentally through reduced pesticide usage.

Annabel et al. [13] noted that machine learning algorithms that use classification to detect plant leaf diseases can be beneficial as they can cut down on expenses of having human workers detect it themselves. The potential setback to this is that the accuracy of the machine learning algorithms can fluctuate between different types of plants, so it is important for them to be trained with a variety of datasets from different types of plants.

In 2018, Kulkarni [38] discussed the use of Convolutional Neural Networks (CNNs) for crop management and how they can specifically be used for crop and disease identification. Kulkarni [38] continued by stating that to achieve this, the CNN breaks the process down into two steps by first determining the crop species and then the type of disease that may be present on it.

More recent applications of CNNs in agriculture include detecting weeds in strawberry fields [39], potato fields [40], wild blueberry fields [41,42], and Florida vegetables [43]. They have also been utilized for detecting diseases on tomato [44,45], apple, strawberry, and various other plants [45].

Farmers have been using herbicides in their field for weed management as it provides benefits in the form of reduced production and labor costs and an increase in crop yields [46]. The use of herbicides does come at an environmental cost as excessive use can contaminate nearby water sources [46,47]. The solution to this issue is to change the method in which herbicides are sprayed in the field. Rather than spray entire acres of crop fields, researchers have focused on creating technology that would specifically tell the farmer where the herbicide must be sprayed.

9.4.2 Livestock management

In 2016, Morales et al. [48] conducted research using an SVM for livestock production, specifically for analyzing the production curve of commercial eggs and having it warn the user if any diseases arise in the laying hens. They determined that their SVM could predict a drop in egg production up to 3 days in advance of experts. Similarly, in 2017, Matthews et al. [49] researched the use of machine learning for animal welfare. In the paper, they focused on the issue of monitoring animals in a commercial setting and the impracticality of manual assessment given the financial and time-associated costs. The solution to this issue was to implement a machine learning system that would automate the process by analyzing the behavior of pigs. For this automated monitoring system, GMMs are used.

In 2018, Hansen et al. [50] conducted research on the use of CNNs for livestock production. The focus of their research was an alternative to current practices for tagging pigs in a commercial setting. The current practice of identifying livestock such as cows and pigs involves tagging the animal in the form of piercing which can be intense for them. The alternative that they propose is using CNNs in a facial recognition system that will allow for farmers to easily identify pigs without the time and cost of tagging them.

9.4.3 Soil management

The review by Liakos et al. [14] found that there are four different properties that researchers have used machine learning algorithms to improve upon. The four properties are soil drainage, soil conditions, soil temperature, and soil moisture.

Soil drainage is an important property for farmers to know, so that they can have proper control of soil and moisture management on their farm. In 2014, Coopersmith et al. [51] assessed the potential for machine learning to be used for evaluating soil drainage in field. In the paper, they used classification trees, a kNN algorithm, and a boosted performance algorithm for their research. It was determined that out of the three machine learning algorithms used, classification trees ranked the lowest on performance while the kNN algorithm performed slightly better than the boosted performance algorithm.

In 2016, Johann et al. [52] researched the use of machine learning algorithms for estimating the soil moisture in fields. For the research, two different ANN algorithms were used and compared to different multiple linear regression models. It was found that both ANN algorithms are viable for the estimation of soil moisture and that they carried performance advantages over the different multiple linear regression models such as simplicity.

9.4.4 Water management

Liakos et al. [14] states that the two main properties of water management that machine learning algorithms can improve are evapotranspiration and daily dew point temperature. Additionally, machine learning has been used to assess water quality in agriculture [47] Machine learning algorithms can be designed for weather predictions in the field by using available weather data [53]. This weather data can include climate records and satellite imagery. The use of sensors in the field can be beneficial as well for providing current data that will allow for greater accuracy of climatic prediction.

In 2017, Cramer et al. [54] conducted research on the use of several machine learning algorithms to accurately predict rainfall. The purpose of the research was to improve the accuracy of information given to farmers, so that they can plan around issues such as drought or flooding. For crop farmers, information on the amount of rainfall for any given season is vital since they financially depend on the crops in their field to get the correct amount of water. Notable machine learning algorithms that are used in the paper are kNN, Support Vector Regression, Genetic Programming, and Radial Basis Neural Networks.

In 2015, Mohammadi et al. [55] researched the use of an Extreme Learning Machine (ELM) for predicting the daily dew point temperature. In the paper, they compare the accuracy and reliability of an ELM against other machine learning algorithms such as SVM and ANN. The accuracy of predicting the daily dew point temperature in a field is important, as it can be an indicator for moisture in the air and for potential rain or snowfall. It was determined in the paper that the ELM performed the prediction quicker and more accurately than the SVM and ANN used.

In 2020, Chen et al. [47] modified the pooling layer of a CNN with a decision tree for detecting water pollution near-infrared data. The authors used the modified CNN to predict the level of chemical oxygen demand (COD), the quantity of oxygen needed to treat harmful pollutants, in agricultural irrigation resources.

9.5 Case study: deep neural networks for ripeness detection in wild blueberry

Wild blueberries (*Vaccinium angustifolium* Ait.) are a perennial crop native to northeastern North America. Unlike traditional crops, they cannot be seeded but rather are encouraged to grow from existing rhizomes within the soil through the removal of competing vegetation [56]. To maximize fruit production, wild blueberries are typically managed on a 2-year cycle which begins with shoot emergence in late May and continued growth of the plant stem until tip dieback occurs toward the end of July. From this point, the blueberry plant will see very little in the way of further vertical growth reaching an average height around 30 cm [57]. The plant then overwinters, and the second season sees flower bud emergence with full flowering occurring in late May to early June. Pollinated flowers will then develop berries which begin green and transition to pink, red, and finally a blue or black color as they ripen. Fully ripened berries can range in size from 0.48 to 1.27 cm diameter [57,58].

Harvest timing is one of the critical decision points in the production cycle of wild blueberries due to the small window that exists when berries are at their peak ripeness. Harvesting during this approximately 5-day period ensures that minimal damage is incurred by the blueberries when they are harvested. The trouble which arises, is determining when this period will occur, can be a challenge and as all fields do not ripen at the same time, there is the potential to develop a harvest plan which capitalizes on this temporal discrepancy. This is where neural networks can play a role in helping growers to better scout their fields and understand how their different fields are developing in relation to one another.

As part of an independent study, MacEachern et al. [59] trained a series of CNNs which could detect wild blueberries and classify them by ripeness. They began by collecting a series of 864 images at 6000 × 4000 pixel resolution across four wild blueberry fields in 2018 and 2019. These 864 images were cropped into 17,280 subimages with dimensions of 1280 × 720 pixels. From these 17,280 subimages, 6766 were randomly selected for CNN model development. Of these images, 70% were used for model training, while 15% were reserved for validation, and a further 15% were reserved for independent model testing. Each of the 6766 randomly selected images had their berries labeled as either green (unripe), red (unripe), or blue (ripe) by drawing a box around them using a custom-built software. The bounding boxes allow the CNN models to learn by providing them with a series of reference points for each of the ripeness classes.

A selection of four different networks from the YOLO (You Only Look Once) family of CNNs [60] was used to develop the models. These included the base network YOLOv3 [61], the smaller YOLOv3-Tiny [62], the enhanced YOLOv3-SPP [63] which includes a spatial pyramid pooling layer, and the modern YOLOv4 [64]. Each of the models were trained using an NVIDIA Quadro RTX 5000 Max-Q 16 GB GPU on the Windows 10 Pro operating system.

YOLOv3, YOLOv3-SPP, and YOLOv4 each returned similar F1-scores of 0.77, 0.77, and 0.75, respectively. That said, when attempting to provide real-time ripeness analysis for growers, inference time is of equal importance. Across the four models, inference times achieved using 1280×720 images were 22.88 ms, 23.15 ms, 5.51 ms and 33.90 ms for each of YOLOv3, YOLOv3-SPP, YOLOv3-Tiny and YOLOv4, respectively. Comparatively, the video random access memory (VRAM) usage was analyzed using the same 1280×720 images (Table.9.2).

It was determined that YOLOv3 represented the best combination of accuracy, inference time, and memory use. Even though YOLOv3-Tiny saw a far shorter inference time, its substantially lower F1-score ultimately led to it being a suboptimal choice. Future work will involve incorporating yield data into the models while developing a mobile app for use on smartphones in field. This would allow growers to estimate potential yields from a series of images in real time on their personal devices. Further, it would afford growers the ability to better determine ripeness stages across all their fields and develop harvest plans, which better compensate for the differing ripening conditions at each site.

Ripeness detection using neural networks has likewise been carried out in several other cropping systems. Mazen and Nashat [65] developed a model for analyzing banana ripeness which returned detection accuracies of 100% for unripe green and overripe brown bananas and 97.75% for ripe yellow bananas in a testing set of 89 images. Despite these high accuracies, it should be noted that these models were tested in a laboratory setting with bananas placed against a white background. In field ripeness determinations would introduce a significant amount of background

Table 9.2 F1-score, inference time, and VRAM usage across the four trained models when analyzing a single, independent, 1280×720 pixel image.

Model	F1-score	Inference time (ms)	VRAM (MB)
YOLOv3	0.78	22.88	1396
YOLOv3-SPP	0.78	23.15	1417
YOLOv3-Tiny	0.67	5.51	635
YOLOv4	0.76	33.90	2317

noise which would likely decrease the accuracies of any developed model. In looking at a study more similar to what was done by MacEachern et al. [66], Kangune et al. [67] analyzed grape ripeness on vines. Their study analyzed 4000 images of both ripe and unripe grapes and classified them accordingly. Their study returned maximum accuracies of 79% when run as an image classifier. This is an important distinction as image classifiers make inference about the entire image rather than attempting to localize objects within images. Therefore, when shown an image, their model simply returns ripe or unripe and cannot make inference about individual fruit as was done with the MacEachern et al. [66] study. By making inference about individual berries, growers can better determine the ripeness stage of their fields overall. This is particularly important in wild blueberries as clonal variations within the species can result in drastically different ripening patterns. This increased level of detail does come at the cost of better accuracies possible with object classifiers along with increased inference time. Determining which machine learning approach is better would need to be performed on a case-by-case basis.

9.6 Applications in precision agriculture

As technology developments such as Global Navigation Satellite System (GNSS) and Geographic Information Systems (GIS) advance along with the development of new sensors, so too does precision agriculture technology. With these developing systems, the social, environmental, and economic sustainability and competitiveness of precision agriculture likewise increases (Perez-Ruiz et al., 2012). Many robotic systems reduce the need for human labor while increasing yield and reducing inputs. Those technologies at the forefront of the precision agriculture revolution include autoguidance and autosteering systems, uncrewed aerial vehicles, robotics, and variable rate agrochemical applications.

9.6.1 Autoguidance

Autoguidance or autosteer is a system that moves a piece of equipment along a set path without the input of the operator, through the employment of GNSS. Depending on the autosteer system, this can include corrections from one or more satellite constellations including GPS, Galileo, Beidou, or GLONASS (Globalnaya Navigazionnaya Sputnikovaya Sistema, or Global Navigation Satellite System). Some systems, such as John Deere's Starfire Series, also include a global network of ground based correction signals to further enhance system accuracies [68]. Beyond the correction signals, it is equally important that the piece of equipment is properly defined, so that the system knows exactly where the GNSS receiver and working point of any implements are spatially located with respect to one another. Finally, there are two primary methods through which autosteering is achieved. The first option is through the employment of a high torque electric motor, which attaches directly onto the steering column of the tractor. These systems are far easier to move between machinery but do come at the cost of slight accuracy and precision reductions. The alternative is autosteering systems which plumb directly into the hydraulics of the machinery. These systems are more precise than bolt-on options but are far more difficult and time consuming to move between equipment. Regardless of the system, it has been thoroughly concluded that GNSS-based autosteering systems are a wise investment for farmers from a profitability and efficiency standpoint [69–71]. Several studies have gone on to demonstrate the benefits of GNSS autosteer compared with a skilled operator in a variety of cropping systems. Lipiński et al., 2016 [72] demonstrated a reduction in tillage overlap from 277 mm with a skilled operator to only 16 mm with an autosteer system. Samenko et al., 2020 [73] showed that peanut harvester deviation reduced from a maximum of 177 mm with a skilled operator to only 11.4 mm with autosteer. Finally, Esau et al. (2021) [74] showed that even at speeds as low as 0.31 m/s autosteer was able to reduce absolute discrepancy from 113.2 to 73.8 mm when harvesting wild blueberries. Low speed autoguidance is a practice which has previously been impossible; however, with advancements in GPS accuracies and auto steering systems, it is now a reality.

Autoguidance of agricultural equipment has likewise been performed through the employment of machine vision. Various studies have looked at automating agricultural equipment in row crops [75–81], cereals [82,83], rice [84], and tree crops [85] though the technology has largely stagnated in recent years. This is primarily due to the fact that these systems have proven to be more unreliable than, and lack the precision and accuracy of, their GNSS-based autoguidance counterparts [86].

9.6.2 Uncrewed aerial vehicles

Uncrewed Aerial Vehicles (UAV or drone) have become one of the more commonly employed precision agriculture systems in recent years. UAVs offer a wide variety of management options across various cropping and livestock operations. That said UAVs are most often utilized in remote sensing applications by pairing them with an appropriate sensor. Some of the more commonly employed sensors include passive sensors such as RGB cameras, thermal cameras, multispectral and hyperspectral cameras, as well as active sensors like synthetic aperture radar and Light Detection and Ranging (LiDAR). Regardless of the sensor, UAVs provide an aerial perspective on one's operation, and from that perspective, valuable information can be ascertained. One of the more prevalent metrics used in modern crop agriculture are remote sensing indices, developed from multi- and hyperspectral imagery. These indices allow the visualization of complex phenomena based on the reflectance characteristics within an image. Normalized Difference Vegetation Index (NDVI) is the most widely employed vegetation index in crop agriculture [87,88]. NDVI is best employed as tool for monitoring plant vigor and canopy health within a crop. NDVI measurements are performed at the pixel level of an image and then averaged across a ground truthed area for interpolation [89]. Values for NDVI will range from -1 to 1 with -1 being water, 0 being bare soil or rocks and 1 being highly vigorous, dense vegetation. NDVI's greatest drawback is its inability to differentiate between good and great plant vigor. For this reason, its value is reduced in well-established fields or for late season observations in successful crops. That said, there are many other vegetation indices which can accomplish this and other observations with a list of over 500 being available from The IDB Project [90].

Remote sampling is another recently adapted usage for UAVs. Employment of UAVs in this fashion has the potential to significantly speed up the ground truthing requirement often associated with various drone-based imaging tasks [91]. Further, drone-based water, air, soil, and tissue sample collection are all seeing increased research, and could become a commercial reality in the years to come [92–95].

Currently, Canadian regulations do not allow the spraying of agrochemical from a UAV-based platform. With that said countries such as the United States, South Korea, Japan, and Australia have all demonstrated that it can be done safely and with great success. Drone-based spraying offers a number of unique advantages over conventional approaches including reduced agrochemical usage, reduced labor requirement, improved spray uniformity, reduced soil compaction, reduced environmental contamination potential and improved safety for operators when considering chemical exposure and machinery injury potential [96]. Today, commercially available options such as the DJI AGRAS T30 (SZ DJI Technology Co., Ltd., Nanshan, Shenzhen, China) offer a 30 L spray tank with 16 nozzles capable of spraying over 16 ha per hour. The system can also be adapted to accommodate granular applications with a 40 L tank capacity and 7 m spread width. In Canada, it is only government regulation which is standing in the way of employing drone-based agrochemical application, a technique which will improve yields and profits while simultaneously improving social and environmental sustainability.

9.6.3 Ground robots

Ground-based robotics can refer to either self-propelled robots or robotic implements that are transported via a vehicle or piece of machinery. Ground robots have the potential to greatly improve the efficiency of many agricultural tasks; however, commercially available options are limited [97]. There are several key obstacles impeding the development of agricultural robotics with the two most prevalent being navigation and sensing system accuracy. Navigation of agricultural robots has seen its greatest success through the employment of GNSS-based navigation [98]. That said, obstacle avoidance approaches including vision (color cameras), infrared, ultrasonic, and laser sensors have likewise been employed to varying degrees of success [99]. Further techniques such as rails and buried guidance lines have also been used [99,100], though these systems rely on additional infrastructure which is costly, requires upkeep, and could impede other agricultural equipment.

For self-guided robots, the most common mechanisms used to maneuver are wheels and crawler tracks [100]. Wheels are often chosen due to their simplicity but can struggle in areas of soft, wet, or uneven soil where the robots can become immobilized. Crawler tracks account for this well but are a more complex design and often require lower travel speeds than wheel-based alternatives. Crawler designs also provide a more stable platform which can be beneficial dependent on the employed task of the robot [100,101].

Agricultural robots have seen usage in a wide variety of tasks including crop harvesting [101–103], weeding [104–108], spraying [109–113], mowing [114], planting/seeding [115–118], and irrigation [119–122]. That said specialized deployment of robots is only possible through pairing robotics with other technologies such as GNSS, machine vision, and other sensors. For example, fruit harvesting robots can be paired with machine vision to precisely identify where fruit is at on a tree and pick it accordingly [123]. Robotic planters such as the Fendt Xaver or MARS (Mobile Agriculture Robot Swarm) are reliant on GNSS to achieve their function. When compared with conventional tractor-based approaches, robotics offers some further advantages. First is the potential for agricultural robots to significantly reduce the effect of soil compaction. Being that agricultural robots are typically smaller and lighter than their tractor counterparts, they exhibit far less impact on croplands. Second is the ability for robots to operate 24 hour per day. As robots do not encounter the same challenges as humans (need for sleep/rest and nighttime visibility) there exists the opportunity to drastically improve efficiencies through robotics. Likewise, we know that there is a definitive link between human fatigue and operational safety, especially with tasks of monotonous or repetitive nature [124]. Many agriculture tasks fit this description and robots offer a greater sense of safety in that respect. Finally, robots are often more environmentally sustainable as they do not consume the substantial amount of fossil fuels typical of a fleet of tractors. Many robots operate on a battery or use renewable energy sources such as solar for achieving their goals [125–127].

9.6.4 Precision irrigation

Currently, irrigation practices account for 70% of the world's fresh consumption with this number rising to 80% in developing countries [128,129]. Approximately 40% of global food production is reliant on irrigation with East Asia comprising of over 70% of the world's irrigated cropland [130]. Compounding this is the reduced irrigation efficiency observed in East Asia, where countries like India have been shown to utilize up to four times as much irrigation water per unit of food as North American and European irrigation systems [131]. With global freshwater resources already in short supply, and the estimate that only half of all irrigation water actually reaches the intended crop, the need for improved irrigations systems is paramount [129,132,133].

Precision irrigation refers to more than simply reducing cropland freshwater inputs. It needs to consider the spatiotemporal variability within a field and provide irrigation water to account for varying crop-specific deficiencies across that field. There are various methods employed for understanding the water requirements within a field but what's common to all fields is soil. Most simply, soil moisture monitoring can be performed using soil probes. While often being an affordable option, soil probes are limited to the immediate area they are placed in. However, connecting a series of probes into a network allows for real-time monitoring of soil moisture with the potential benefits of monitoring other key soil parameters such as temperature, pH, and salinity [134]. That said, depending on the size of the operation, this may not be an economically feasible option. An alternative is to utilize UAVs paired with a multi/hyperspectral camera. By applying an appropriate index such as NDVI, NDWI (Normalized Difference Water Index), or PDI (Perpendicular Drought Index) to a multispectral image, one can monitor soil moisture within a field [132,135,136]. Moreover, application of indices can be paired with machine learning to further enhance the potential for identifying trends or problem areas within a field [136]. Another indirect method for monitoring soil moisture is through infrared thermometry. By incorporating such a sensor on a UAV and generating a temperature map of a field, one can indirectly monitor the soil moisture in a monocrop [137]. This is because a plant with ample access to water will readily transpire during the heat of the day while a plant with a water deficit will transpire far less. The transpiration process cools the plant and this gradient can be mapped across a field to identify problem areas [138]. The greatest challenge with UAV-based approaches is that they require time to fly, process and produce the data. Compared with soil probes, their time to action is drastically increased however, once the data is organized, they offer much greater resolution and potential for improved management decisions.

The simplest form of precision irrigation is through drip irrigation. This approach drastically cuts back on freshwater usage by delivering water directly to plant roots. The best application of drip irrigation is for irrigation scheduling in largely homogenous field conditions [139]. That said, varying flow rates within a field is often not possible and the nature of these systems in an outdoor environment means maintenance requirements are high. For these reasons, precision irrigation is typically performed using continuous move irrigators such as center pivot, linear move, and boom and reel irrigators [132]. By combining a prescription map with pulse width modulated nozzles, various researches have demonstrated the potential for semi-automated spot-specific irrigation [140–146]. Implementation of this process does however remain quite time consuming and finding methods to replace the prescription mapping requirement should be at the forefront of irrigation research.

9.6.5 Variable rate and spot specific agrochemical application

Variable rate agrochemical applications have been employed since the 1980s [147]. The years since have seen considerable advancements in this area, traditionally through the employment of prescription maps, and now through machine vision—based approaches. Prescription maps can be developed in a variety of ways with the simplest being a manual approach, where an individual walks a field and georeference unique points of interest. Evidently, this approach can be quite time-consuming depending on the level of precision required from the produced map. That said, predictive algorithms such as kriging, can be employed to help reduce the data collection requirement [148]. In this approach, a limited number of datapoints are collected and based on the collected values and their proximity to one another, the rest of the map can be extrapolated. However, Heisel et al. [148] went on to note that despite the potential of kriging, their analysis still required far too many datapoints for practical application. The next step in prescription map production is therefore to mount data collection equipment onto a vehicle. The first adaptation of this technique utilized a push button system which georeferenced its location based on a wheel sensor and a known start location in a field row [149]. By logging the button presses one could develop a simple prescription map of the target and field, significantly speeding up the collection process at the cost of accuracy. The advent of UAVs and their eventual civilian legality is where the potential of prescription mapping has been truly realized. UAVs can be paired with a range of sensors including color cameras, multi/hyperspectral cameras, thermal cameras, LiDAR, and synthetic aperture radar, to generate maps with varying parameters. Through an aerial perspective and the high levels of accuracy and precision achievable with modern sensors, Uncrewed Aerial Systems (UAS) are able to produce georeferenced datasets far more effectively than traditional approaches. UASs have the further benefit of being nondestructive, and unlike similar satellite or crewed aircraft—based approaches, offer far greater resolutions due to their ability to fly at low altitude [150–152]. Regardless of how the maps are generated, they need to be paired with an appropriate rate controller to regulate application rate and management zones. Several companies produce such rate controllers with some of the more common being outlined in Table 9.3.

Despite the plethora of options available, a common problem remains with prescription mapping, that being the time it takes to produce a map. Reliance on prescription mapping is far from a real-time solution, and for an industry which is often already stretched for time, it is not an optimized solution.

Machine vision—based approaches are a real-time solution to precision agrochemical application. Research and industry have explored different options for achieving this goal with a few solutions now being commercially available in the form of the John Deere See & Spray and the Agrifac WEED-IT (EXEL Industries, Épernay, FR). These systems both rely on the reflectance characteristics of their

Table 9.3 List of common rate controllers used in combination with prescription maps.

Company	Location	Rate controller
John Deere	Moline IL, US	RC2000
CHN industrial	Amsterdam, NL	Raven rate control module
Trimble	Sunnyvale CA, US	Field IQ
AGCO	Duluth GA, US	Fuse rate and section controller
Fendt	Marktoberdorf, DE	Variotronic (integrated)

target by seeking out green color and spraying accordingly. While this approach has potential in cereal crops or against bare soil, it will struggle when determining between green weeds and green plants. This is where neural network—based machine vision systems come in. Neural network systems are not reliant on color alone; rather they analyze images on a pixel level and look for patterns within the image to identify the target. Training neural networks is performed through backpropagation, where labeled images are shown to the network and the patterns can be identified after hundreds or thousands of iterations [23]. In this way, a sprayer fitted with a neural network—based machine vision system can drive through a field, determine where the targets are, and spray them accordingly. Several networks have already been developed in the agricultural sector for identifying a variety of weeds, diseases, and crops; these can be observed in Table 9.4.

To date, there has been some success in the development of precision spot applicators based on machine vision. Alam et al. [163] developed a four-wheeled robot, which distinguishes between tobacco plants and weeds and sprays the weeds accordingly. Their design resulted in a 52% reduction in pesticide usage when compared with traditional practices. Liu et al. [164], developed a similar four-wheeled robot which could identify weeds among strawberry plants and apply herbicide accordingly. Hussain et al. [40] developed a ride on sprayer buggy which detects weeds between potato rows and applies herbicide. In testing, their buggy was able to save up to 43% more herbicide than traditional applications while being 90% accurate at detecting target weeds. Partel et al. [46] developed an ATV mounted system which

Table 9.4 Sample of weed and disease detection algorithms.

Authors	Application
Zheng et al. [153]	30 crop specific classes
Cap et al. [154]	6 food crops and 8 weeds
Olsen et al. [155]	8 unique weeds
Champ et al. [156]	4 weeds in beans and corn
Lameski et al. [157]	Weeds in carrots
Al-Qurran et al. [158]	5 unique leaf diseases
Giselsson et al. [159]	14 crops and weeds
Espejo-Garcia et al. [160]	Common weeds in tomato and cotton
Hennessy et al. [41]	2 weeds in wild blueberry
MacEachern et al. [59]	Wild blueberry ripeness stage
Sharpe et al. [39]	Weeds in strawberry
Hussain et al. [40]	Weeds and blight in potato
Yu et al. [161]	Broadleaf weeds in turfgrass
Sharpe et al. [43]	Vegetation classifiers in vegetable crops
Fuentes et al. [44]	9 leaf diseases in tomato
Venkataramanan et al. [45]	Leaf diseases in 8 crops
Schumann et al. [162]	Citrus leaf diseases

targets weeds in field crops with an average precision and recall of 71% and 78%, respectively. Finally, Partel et al. [165] also developed a machine vision–based tree sprayer which identifies the tree's dimensions, counts the fruit present on the tree, and sprays in the desired areas. In all, the system saves 28% more pesticide than traditional management options while simultaneously providing a rough yield estimation. Systems such as those discussed above are paving the way for future developments in precision agrochemical application. While prescription mapping can produce high accuracy and precision, its deployment is time consuming and real-time applications such as machine vision–based approaches offer considerable potential moving forward.

9.7 Conclusions and future implications

Industry and research are rapidly progressing with agriculture's fourth revolution, characterized by the adoption of big data, the internet of things and artificial intelligence to improve the way crops and livestock are produced. Currently, farmers have access to a plethora of data collection and information processing options to aid them in making decisions about their operations. Despite this, research is lacking in how all this data can be connected and employed in a meaningful way. Artificial intelligence can be the solution to this challenge due to its potential to process large sums of data, and provide logic-based solutions, recommendations, and actions accordingly. Machine learning in particular has the ability to change the way we view farming. Due to the vast data processing capabilities of modern computing hardware, machine learning algorithms can identify previously unexplored patterns within datasets, leading to improvements in current and future practices. Ultimately, the primary goal of modern agriculture is to produce more food with less inputs, all the while dealing with the current constraints on arable farmland, agrochemicals, labor, and freshwater. Artificial intelligence–based and reliant technologies such as GNSS, GIS, UAVs, robotics, and machine vision will be major contributors in how agriculture will look in the coming years. By allowing farmers to make more informed decisions about their operations, while optimizing current practices and reducing manual labor requirements, artificial intelligence will be a critical ally in the push to feed the world.

References

[1] Solomonoff RJ. The time scale of artificial intelligence. Hum Syst Manag 1985;5: 149–53.
[2] McCarthy J, Minsky ML, Rochester N, Shannon C. A proposal for the Dartmouth summer research project on artificial intelligence.27 4th Ed. AI Magazine; 1955, 12.
[3] Mitchell TM. Machine learning. 1st ed. McGraw Hill Higher Education; 1997.

[4] Goodfellow I, Bengio Y, Courville A. Deep learning. 1st ed. Cambridge: The MIT Press; 2016.
[5] Breunig MM, Kriegel H-P, Ng RT, Sander JLOF. Identifying density-based local outliers. In: Proceedings of 2000 ACM SIGMOD ACM international conference on management of data; 2000. p. 93–104. https://doi.org/10.1145/2F335191.335388.
[6] MacQueen J. Some methods for classification and analysis of multivaritate observations. Fifth Berkeley Symp Math Stat Probab 1967;5(1):281–97.
[7] Kaelbling LP, Littman ML, Moore AW. Reinforcement learning: a survey. J Artif Intell Res 1996;4:237–85.
[8] Cortes C, Vapnik V. Support-vector networks. Mach Learn 1995;20:273–97.
[9] Quinlan JR. Induction of decision trees. Mach Learn 1986;1:81–106. https://doi.org/10.1007/bf00116251.
[10] Rivest RL. Learning decision lists. Mach Learn 1987;2:229–46. https://doi.org/10.1023/A:1022607331053.
[11] Fix E, Hodges JL. Discriminatory analysis. Nonparametric discrimination: consistency properties. Texas: Randolph Field; 1951.
[12] Altman NS. An introduction to kernel and nearest-neighbor nonparametric regression. Am Statistician 1992;46:175–85. https://doi.org/10.1080/00031305.1992.10475879.
[13] Annabel LSP, Annapoorani T, Deepalakshmi P. Machine learning for plant leaf disease detection and classification—a review. In: Proceedings of 2019 IEEE international conference on communication and signal processing, ICCSP 2019; 2019. p. 538–42. https://doi.org/10.1109/ICCSP.2019.8698004.
[14] Liakos KG, Busato P, Moshou D, Pearson S, Bochtis D. Machine learning in agriculture: a review. Sensors 2018;18:1–29. https://doi.org/10.3390/s18082674.
[15] Ho TK. Random decision forests. In: Proceedings of international conference on document analysis and recognition, ICDAR. 1; 1995. p. 278–82. https://doi.org/10.1109/ICDAR.1995.598994.
[16] Rosenblatt F. The perceptron: a probabilistic model for information storage and organization in the brain. Psychol Rev 1958;65:386–408. https://doi.org/10.1037/h0042519.
[17] Minsky ML, Papert S. Perceptrons: an introduction to computational geometry. 1st ed. Cambridge: The MIT Press; 1969.
[18] Olazaran M. A sociological study of the official history of the perceptrons controversy. Soc Stud Sci 1996;26:611–59.
[19] Dunn JC. A fuzzy relative of the ISODATA process and its use in detecting compact well-separated clusters. J Cybern 1973;3:32–57. https://doi.org/10.1080/01969727308546046.
[20] Bezdek JC, Ehrlich R, Full W. FCM: the fuzzy c-means clustering algorithm. Comput Geosci 1984;10:191–203.
[21] Rehman TU, Mahmud MS, Chang YK, Jin J, Shin J. Current and future applications of statistical machine learning algorithms for agricultural machine vision systems. Comput Electron Agric 2019;156:585–605. https://doi.org/10.1016/j.compag.2018.12.006.
[22] Reynolds D. Gaussian mixture models. Encycl Biometrics 2009;741:659–63.
[23] Rumelhart D, Hinton G, Williams RJ. Learning representations by back-propagating errors. Nature 1986;323.
[24] Cauchy A-L. Methode generale pour la resolution des systemes d'equations simultanees. Compte Rendu Des Seances L'Acad'emie Des Sci 1847;25:536–8.

[25] Harris M. Many-core GPU computing with NVIDIA CUDA. In: Proceedings of 22nd annual international conference on supercomputing. Vol. 1; 2008. https://doi.org/10.1145/1375527.1375528.
[26] Zeiler MD, Fergus R. Visualizing and understanding convolutional networks. 2013. https://doi.org/10.1111/j.1475-4932.1954.tb03086.x. ArXiv.
[27] Jarrett K, Kavukcuoglu K, Ranzato M, LeCun Y. What is the best multi-stage architecture for object recognition?. In: 2009 IEEE 12th international conference on computer vision, Kyoto; 2009. p. 2146−53. https://doi.org/10.1109/ICCV.2009.5459469.
[28] Nair V, Hinton G. Rectified linear units improve restricted Boltzmann machines. In: Proceedings of 27th international conference on machine learning, Haifa; 2010.
[29] Boureau Y, Ponce J, LeCun Y. A theoretical analysis of feature pooling in visual recognition. In: Proceedings of 27th international conference on machine learning; 2010. p. 111−8.
[30] Zhou YT, Chellappa R. Computation of optical flow using a neural network. In: IEEE international conference on neural networks; 1988. p. 71−8.
[31] LeCun Y, Bottou L, Bengio Y, Haffner P. Gradient-based learning applied to document recognition. Biochem Biophys Res Commun 1998;86:2278−324.
[32] Krizhevsky A, Sutskever I, Hinton G. ImageNet classification with deep convolutional neural networks. 2012.
[33] Russakovsky O, Deng J, Su H, Krause J, Satheesh S, Ma S, et al. ImageNet large scale visual recognition challenge. Int J Comput Vis 2015;115:211−52. https://doi.org/10.1007/s11263-015-0816-y.
[34] Kamilaris A, Prenafeta-Boldú FX. Deep learning in agriculture: a survey. Comput Electron Agric 2018;147:70−90. https://doi.org/10.1016/j.compag.2018.02.016.
[35] Benos L, Tagarakis AC, Dolias G, Berruto R, Kateris D, Bochtis D. Machine learning in agriculture: a comprehensive updated review. Sensors 2021;21:1−55. https://doi.org/10.3390/s21113758.
[36] Ramos PJ, Prieto FA, Montoya EC, Oliveros CE. Automatic fruit count on coffee branches using computer vision. Comput Electron Agric 2017;137:9−22. https://doi.org/10.1016/j.compag.2017.03.010.
[37] Sharma R, Kamble SS, Gunasekaran A, Kumar V, Kumar A. A systematic literature review on machine learning applications for sustainable agriculture supply chain performance. Comput Oper Res 2020;119:104926. https://doi.org/10.1016/j.cor.2020.104926.
[38] Kulkarni O. Crop disease detection using deep learning. In: Fourth international conference on computing communication control and automation; 2018. p. 1797−804.
[39] Sharpe SM, Schumann AW, Boyd NS. Detection of *Carolina geranium* (*Geranium carolinianum*) growing in competition with strawberry using convolutional neural networks. Weed Sci 2019;67:239−45. https://doi.org/10.1017/wsc.2018.66.
[40] Hussain N, Farooque AA, Schumann AW, McKenzie-Gopsill A, Esau T, Abbas F, et al. Design and development of a smart variable rate sprayer using deep learning. Rem Sens 2020;12:4091.
[41] Hennessy PJ, Esau TJ, Farooque AA, Schumann AW, Zaman QU, Corscadden KW. Hair fescue and sheep sorrel identification using deep learning in wild blueberry production. Rem Sens 2021;13:943.
[42] Hennessy PJ, Esau TJ, Schumann AW, Zaman QU, Corscadden KW, Farooque AA. Evaluation of cameras and image distance for CNN-based weed detection in wild blueberry. Smart Agric Technol 2021;2. https://doi.org/10.1016/j.atech.2021.100030.

[43] Sharpe SM, Schumann AW, Yu J, Boyd NS. Vegetation detection and discrimination within vegetable plasticulture row-middles using a convolutional neural network. Precis Agric 2020;21:264—77. https://doi.org/10.1007/s11119-019-09666-6.

[44] Fuentes A, Yoon S, Kim SC, Park DS. A robust deep-learning-based detector for real-time tomato plant diseases and pests recognition. Sensors 2017;17. https://doi.org/10.3390/s17092022.

[45] Venkataramanan A, DKP H, Agarwal P. Plant disease detection and classification using deep neural networks. Int J Comput Sci Eng 2019;11:40—6.

[46] Partel V, Kakarla SC, Ampatzidis Y. Development and evaluation of a low-cost and smart technology for precision weed management utilizing artificial intelligence. Comput Electron Agric 2019;157:339—50.

[47] Chen H, Chen A, Xu L, Xie H, Qiao H, Lin Q, et al. A deep learning CNN architecture applied in smart near-infrared analysis of water pollution for agricultural irrigation resources. Agric Water Manag 2020;240:106303.

[48] Morales IR, Cebrián DR, Fernandez-Blanco E, Sierra AP. Early warning in egg production curves from commercial hens: a SVM approach. Comput Electron Agric 2016;121:169—79. https://doi.org/10.1016/j.compag.2015.12.009.

[49] Matthews SG, Miller AL, Plötz T, Kyriazakis I. Automated tracking to measure behavioural changes in pigs for health and welfare monitoring. Sci Rep 2017;7:1—12. https://doi.org/10.1038/s41598-017-17451-6.

[50] Hansen MF, Smith ML, Smith LN, Salter MG, Baxter EM, Farish M, et al. Towards on-farm pig face recognition using convolutional neural networks. Comput Ind 2018;98:145—52. https://doi.org/10.1016/j.compind.2018.02.016.

[51] Coopersmith EJ, Minsker BS, Wenzel CE, Gilmore BJ. Machine learning assessments of soil drying for agricultural planning. Comput Electron Agric 2014;104:93—104. https://doi.org/10.1016/j.compag.2014.04.004.

[52] Johann AL, de Araújo AG, Delalibera HC, Hirakawa AR. Soil moisture modeling based on stochastic behavior of forces on a no-till chisel opener. Comput Electron Agric 2016;121:420—8. https://doi.org/10.1016/j.compag.2015.12.020.

[53] Sharma A, Jain A, Gupta P, Chowdary V. Machine learning applications for precision agriculture: a comprehensive review. IEEE Access 2021;9:4843—73. https://doi.org/10.1109/ACCESS.2020.3048415.

[54] Cramer S, Kampouridis M, Freitas AA, Alexandridis AK. An extensive evaluation of seven machine learning methods for rainfall prediction in weather derivatives. Expert Syst Appl 2017;85:169—81. https://doi.org/10.1016/j.eswa.2017.05.029.

[55] Mohammadi K, Shamshirband S, Motamedi S, Petković D, Hashim R, Gocic M. Extreme learning machine based prediction of daily dew point temperature. Comput Electron Agric 2015;117:214—25. https://doi.org/10.1016/j.compag.2015.08.008.

[56] Hall I, Hall H, Le A, Nickerson NL, Vander Kloet SP. The biological flora of Canada. I: vaccinium angustifolium ait. In: Sweet LOWBUSH blueberry. Biol flora Canada I vaccinium angustifolium ait. SWEET LOWBUSH BLUEBERRY; 1979.

[57] Farooque AA, Zaman QU, Groulx D, Schumann AW, Yarborough DE, Nguyen-Quang T. Effect of ground speed and header revolutions on the picking efficiency of a commercial wild blueberry harvester. Appl Eng Agric 2014;30:535—46.

[58] Soule HM. Developing a lowbush blueberry harvester. Trans ASAE (Am Soc Agric Eng) 1969;12:127—9.

[59] MacEachern CB, Esau TJ, Schumann AW, Hennessy PJ, Zaman QU. Detection of fruit maturity stage and yield estimation in wild blueberry using deep learning convolutional neural networks. Smart Agric Technol 2023;3:100099.

[60] Redmon J, Divvala S, Girshick R, Farhadi A. You only look once: unified, real-time object detection. 2015. ArXiv.

[61] Redmon J, Farhadi A. YOLOv3: an incremental improvement. 2018.

[62] Redmon JYOLO. Real-time object detection. 2018. https://pjreddie.com/darknet/yolo/ . [Accessed 18 July 2022].

[63] Huang Z, Wang J. DC-SPP-YOLO: dense connection and spatial pyramid pooling based YOLO for object detection. 2020.

[64] Bochkovskiy A, Wang CY, Liao HYM. YOLOv4: optimal speed and accuracy of object detection. 2020. ArXiv.

[65] Mazen FMA, Nashat AA. Ripeness classification of bananas using an artificial neural network. Arabian J Sci Eng 2019;44:6901—10.

[66] MacEachern CB, Esau TJ, Schumann AW, Hennessy PJ, Zaman QU. Deep learning artificial neural networks for detection of fruit maturity stage in wild blueberries. In: 2021 ASABE annual international virtual meeting. American Society of Agricultural and Biological Engineers; 2021. p. 1.

[67] Kangune K, Kulkarni V, Kosamkar P. Grapes ripeness estimation using convolutional neural network and support vector machine. In: 2019 Global conference advanced technologies. IEEE; 2019. p. 1—5. https://doi.org/10.1109/GCAT47503.2019.8978341.

[68] John Deere. StarFire 6000 receiver manual. 2016.

[69] D'Antoni JM, Mishra AK, Joo H. Farmers' perception of precision technology: the case of autosteer adoption by cotton farmers. Comput Electron Agric 2012;87:121—8.

[70] Ortiz BV, Balkcom KB, Duzy L, Van Santen E, Hartzog DL. Evaluation of agronomic and economic benefits of using RTK-GPS-based auto-steer guidance systems for peanut digging operations. Precis Agric 2013;14:357—75.

[71] Shockley JM, Dillon CR, Stombaugh TS. A whole farm analysis of the influence of auto-steer navigation on net returns, risk, and production practices. J Agric Appl Econ 2011;43:57—75.

[72] Lipiński AJ, Markowski P, Lipiński S, Pyra P. Precision of tractor operations with soil cultivation implements using manual and automatic steering modes. Biosyst Eng 2016; 145:22—8.

[73] Samenko LA, Kirk KR, Turner AP, Fogle BB. Yield recovery effects of autosteering in peanut digging. In: 2020 ASABE annual international virtual meeting. American Society of Agricultural and Biological Engineers; 2020. p. 1.

[74] Esau TJ, MacEachern CB, Farooque AA, Zaman QU. Evaluation of autosteer in rough terrain at low ground speed for commercial wild blueberry harvesting. Agronomy 2021;11:384.

[75] Åstrand B, Baerveldt A-J. A vision based row-following system for agricultural field machinery. Mechatronics 2005;15:251—69. https://doi.org/10.1016/j.mechatronics.2004.05.005.

[76] Søgaard HT, Olsen HJ. Crop row detection for cereal grain. In: Precision Agriculture Part 1 Part 2. Paper presented 2nd European conference on precision agriculture, Odense, Denmark, 11—15 July 1999. Sheffield Academic Press; 1999. p. 181—90.

[77] Lang Z. Image processing based automatic steering control in plantation. VDI-Ber 1998;1449:93—8.

[78] Kise M, Zhang Q, Rovira Más F. A stereovision-based crop row detection method for tractor-automated guidance. Biosyst Eng 2005;90:357—67. https://doi.org/10.1016/j.biosystemseng.2004.12.008.

[79] Hague T, Tillett ND. A bandpass filter-based approach to crop row location and tracking. Mechatronics 2001;11:1—12. https://doi.org/10.1016/S0957-4158(00)00003-9.

[80] Tillett ND, Hague T, Miles SJ. Inter-row vision guidance for mechanical weed control in sugar beet. Comput Electron Agric 2002;33:163—77. https://doi.org/10.1016/S0168-1699(02)00005-4.

[81] Okamoto H, Hamada K, Kataoka T, Terawaki M, Hata S. Automatic guidance system with crop row sensor. In: Automation technology for off-road equipment proceedings of 2002 conference. American Society of Agricultural and Biological Engineers; 2002. p. 307.

[82] Benson ER, Reid JF, Zhang Q. Machine vision—based guidance system for an agricultural small—grain harvester. Trans ASAE (Am Soc Agric Eng) 2003;46:1255.

[83] Tillett ND, Hague T. Computer-Vision-based hoe guidance for cereals—an initial trial. J Agric Eng Res 1999;74:225—36. https://doi.org/10.1006/jaer.1999.0458.

[84] Misao Y, Karahashi M. An image processing based automatic steering rice transplanter (II). An Image Process Based Autom Steer Rice Transplanter (II) 2000:1—5.

[85] Subramanian V, Burks TF, Arroyo AA. Development of machine vision and laser radar based autonomous vehicle guidance systems for citrus grove navigation. Comput Electron Agric 2006;53:130—43. https://doi.org/10.1016/j.compag.2006.06.001.

[86] Vrochidou E, Oustadakis D, Kefalas A, Papakostas GA. Computer vision in self-steering tractors. Machines 2022;10:129.

[87] Huang S, Tang L, Hupy JP, Wang Y, Shao G. A commentary review on the use of normalized difference vegetation index (NDVI) in the era of popular remote sensing. J Res 2021;32:1—6.

[88] Karthikeyan L, Chawla I, Mishra AK. A review of remote sensing applications in agriculture for food security: crop growth and yield, irrigation, and crop losses. J Hydrol 2020;586:124905.

[89] González-Jaramillo V, Fries A, Bendix J. AGB estimation in a tropical mountain forest (TMF) by means of RGB and multispectral images using an unmanned aerial vehicle (UAV). Rem Sens 2019;11:1413.

[90] Henrich V, Krauss G, Götze C, Sandow C. Index DataBase. 2022. https://www.indexdatabase.de/db/i.php. [Accessed 13 June 2022].

[91] van der Merwe D, Burchfield DR, Witt TD, Price KP, Sharda A. Drones in agriculture. Adv Agron 2020;162:1—30.

[92] Lally HT, O'Connor I, Jensen OP, Graham CT. Can drones be used to conduct water sampling in aquatic environments? A review. Sci Total Environ 2019;670:569—75.

[93] Huuskonen J, Oksanen T. Soil sampling with drones and augmented reality in precision agriculture. Comput Electron Agric 2018;154:25—35.

[94] Ruiz-Jimenez J, Zanca N, Lan H, Jussila M, Hartonen K, Riekkola M-L. Aerial drone as a carrier for miniaturized air sampling systems. J Chromatogr A 2019;1597:202—8.

[95] Bieber P, Seifried TM, Burkart J, Gratzl J, Kasper-Giebl A, Schmale DG, et al. A drone-based bioaerosol sampling system to monitor ice nucleation particles in the lower atmosphere. Rem Sens 2020;12:552.

[96] Hafeez A, Husain MA, Singh SP, Chauhan A, Khan MT, Kumar N, et al. Implementation of drone technology for farm monitoring & pesticide spraying: a review. Inf Process Agric 2022.
[97] Santos LC, Santos FN, Pires EJS, Valente A, Costa P, Magalhães S. Path planning for ground robots in agriculture: a short review. In: 2020 IEEE international conference on autonomous robot systems and competitions. IEEE; 2020. p. 61–6.
[98] Vougioukas SG. Agricultural robotics. Annu Rev Control Robot Auton Syst 2019;2: 365–92.
[99] Gul F, Rahiman W, Nazli Alhady SS. A comprehensive study for robot navigation techniques. Cogent Eng 2019;6:1632046.
[100] Roldán JJ, del Cerro J, Garzón-Ramos D, Garcia-Aunon P, Garzón M, De León J, et al. Robots in agriculture: state of art and practical experiences. Serv Robot 2018:67–90.
[101] Chatzimichali AP, Georgilas IP, Tourassis VD. Design of an advanced prototype robot for white asparagus harvesting. In: 2009 IEEE/ASME international conference on advanced intelligent mechatronics. IEEE; 2009. p. 887–92.
[102] Lehnert C, McCool C, Sa I, Perez T. Performance improvements of a sweet pepper harvesting robot in protected cropping environments. J Field Robot 2020;37: 1197–223.
[103] Ling X, Zhao Y, Gong L, Liu C, Wang T. Dual-arm cooperation and implementing for robotic harvesting tomato using binocular vision. Robot Autonom Syst 2019;114: 134–43.
[104] Gokul S, Dhiksith R, Sundaresh SA, Gopinath M. Gesture controlled wireless agricultural weeding robot. In: 2019 5th international conference on advanced computing and communication systems. IEEE; 2019. p. 926–9.
[105] Chen L, Karkee M, He L, Wei Y, Zhang Q. Evaluation of a leveling system for a weeding robot under field condition. IFAC-PapersOnLine 2018;51:368–73.
[106] Lysakov AA, Masyutina GV, Rostova AT, Eliseeva AA, Lubentsov VF. Development of a weeding robot with tubular linear electric motors. In: IOP conference series: earth and environmental science. vol. 852. IOP Publishing; 2021. p. 12063.
[107] Harders LO, Czymmek V, Knoll FJ, Hussmann S. Area yield performance evaluation of a nonchemical weeding robot in organic farming. In: 2021 IEEE international instrumentation and measurement technology conference. IEEE; 2021. p. 1–6.
[108] Mary MF, Yogaraman D. Neural network based weeding robot for crop and weed discrimination. J. Phys. Conf. Ser. 2021;1979:12027. IOP Publishing.
[109] Danton A, Roux J-C, Dance B, Cariou C, Lenain R. Development of a spraying robot for precision agriculture: an edge following approach. In: 2020 IEEE conference on control technology and applications. IEEE; 2020. p. 267–72.
[110] Ghafar ASA, Hajjaj SSH, Gsangaya KR, Sultan MTH, Mail MF, Hua LS. Design and development of a robot for spraying fertilizers and pesticides for agriculture. Mater Today Proc 2021.
[111] Chaitanya P, Kotte D, Srinath A, Kalyan KB. Development of smart pesticide spraying robot. Int J Recent Technol Eng 2020;8:2193–202.
[112] Chrysoulakis C, Fasoulas J, Sfakiotakis M. Development and initial evaluation of a multi-purpose spraying robot prototype. In: 2021 20th international conference on advanced robotics. IEEE; 2021. p. 384–9.
[113] Bhattacharyya N. Design and development of intelligent pesticide spraying system for agricultural robot. In: Hybrid intelligent systems 20th international conference on

hybrid intelligent systems (HIS 2020), December 14–16, 2020, vol. 1375. Springer Nature; 2021. p. 157.
[114] Toyama K, Hatiya K, Koide S, Orikasa T, Shono H, Takeda J. P-BR1: development on the autonomous mowing robot for orchard. J Korean Soc Agric Mach 2018;23:152.
[115] Haibo L, Shuliang D, Zunmin L, Chuijie Y. Study and experiment on a wheat precision seeding robot. J Robot 2015;2015.
[116] Katupitiya J. An autonomous seeder for broad acre crops. In: Proceedings in American society of agricultural and biological engineers annual international meeting; 2014. p. 169–76.
[117] Fendt. Project Xaver: research in the field of agricultural robotics. In: Precision farming—thinking ahead; 2022. https://www.fendt.com/int/xaver. [Accessed 8 June 2022].
[118] Fendt. MARS: robot system for planting and accurate documentation. 2022. https://www.fendt.com/int/fendt-mars. [Accessed 8 June 2022].
[119] Chang C-L, Lin K-M. Smart agricultural machine with a computer vision-based weeding and variable-rate irrigation scheme. Robotics 2018;7:38.
[120] Adeodu AO, Bodunde OP, Daniyan IA, Omitola OO, Akinyoola JO, Adie UC. Development of an autonomous mobile plant irrigation robot for semi structured environment. Procedia Manuf 2019;35:9–15.
[121] Hassan A, Asif RM, Rehman AU, Nishtar Z, Kaabar MKA, Afsar K. Design and development of an irrigation mobile robot. IAES Int J Rob Autom 2021;10:75.
[122] Hassan A, Abdullah HM, Farooq U, Shahzad A, Asif RM, Haider F, et al. A wirelessly controlled robot-based smart irrigation system by exploiting arduino. J Robot Control 2021;2:29–34.
[123] Kuznetsova A, Maleva T, Soloviev V. Using YOLOv3 algorithm with pre-and post-processing for apple detection in fruit-harvesting robot. Agronomy 2020;10:1016.
[124] Williamson A, Lombardi DA, Folkard S, Stutts J, Courtney TK, Connor JL. The link between fatigue and safety. Accid Anal Prev 2011;43:498–515.
[125] Quaglia G, Visconte C, Scimmi LS, Melchiorre M, Cavallone P, Pastorelli S. Design of a UGV powered by solar energy for precision agriculture. Robotics 2020;9:13.
[126] Jothimurugan P, Saravanan JM, Sushanth R, Suresh V, Subramaniam HS, Vasantharaj S, et al. Solar E-Bot for agriculture. In: 2013 Texas instruments India educational conference. IEEE; 2013. p. 125–30.
[127] Plonski PA, Tokekar P, Isler V. Energy-efficient path planning for solar-powered mobile robots. J Field Robot 2013;30:583–601.
[128] Knox JW, Kay MG, Weatherhead EK. Water regulation, crop production, and agricultural water management—understanding farmer perspectives on irrigation efficiency. Agric Water Manag 2012;108:3–8.
[129] Hedley CB, Knox JW, Raine SR, Smith R. Water: advanced irrigation technologies. 2014.
[130] Turral H, Svendsen M, Faures JM. Investing in irrigation: reviewing the past and looking to the future. Agric Water Manag 2010;97:551–60.
[131] Sarma A. Precision irrigation-a tool for sustainable management of irrigation water. In: Proceedings of civil engineering sustainable development challenges, Guwahati, India; 2016. p. 19–21.
[132] Adeyemi O, Grove I, Peets S, Norton T. Advanced monitoring and management systems for improving sustainability in precision irrigation. Sustainability 2017;9:353.

[133] De Fraiture C, Wichelns D. Satisfying future water demands for agriculture. Agric Water Manag 2010;97:502—11.

[134] Abioye EA, Abidin MSZ, Mahmud MSA, Buyamin S, Ishak MHI, Rahman MKIA, et al. A review on monitoring and advanced control strategies for precision irrigation. Comput Electron Agric 2020;173:105441. https://doi.org/10.1016/j.compag.2020.105441.

[135] Casamitjana M, Torres-Madroñero MC, Bernal-Riobo J, Varga D. Soil moisture analysis by means of multispectral images according to land use and spatial resolution on Andosols in the Colombian Andes. Appl Sci 2020;10:5540.

[136] Ge X, Wang J, Ding J, Cao X, Zhang Z, Liu J, et al. Combining UAV-based hyperspectral imagery and machine learning algorithms for soil moisture content monitoring. PeerJ 2019;7:e6926.

[137] Jones HG, Leinonen I. Thermal imaging for the study of plant water relations. J Agric Meteorol 2003;59:205—17.

[138] Jackson RD, Idso SB, Reginato RJ, Pinter Jr PJ. Canopy temperature as a crop water stress indicator. Water Resour Res 1981;17:1133—8.

[139] Mousa AK, Croock MS, Abdullah MN. Fuzzy based decision support model for irrigation system management. Int J Comput Appl 2014;104.

[140] Evans RG, Iversen WM, Kim Y. Integrated decision support, sensor networks, and adaptive control for wireless site-specific sprinkler irrigation. Appl Eng Agric 2011;28:377—87.

[141] Daccache A, Knox JW, Weatherhead EK, Daneshkhah A, Hess TM. Implementing precision irrigation in a humid climate—Recent experiences and on-going challenges. Agric Water Manag 2015;147:135—43.

[142] King BA, Wall RW, Kincaid DC, Westermann DT. Field testing of a variable rate sprinkler and control system for site-specific water and nutrient application. Appl Eng Agric 2005;21:847—53.

[143] Pierce FJ. Precision irrigation. Landbauforsch SH 2010;340:45—56.

[144] Miranda FR, Yoder RE, Wilkerson JB, Odhiambo LO. An autonomous controller for site-specific management of fixed irrigation systems. Comput Electron Agric 2005;48:183—97.

[145] Goumopoulos C, O'Flynn B, Kameas A. Automated zone-specific irrigation with wireless sensor/actuator network and adaptable decision support. Comput Electron Agric 2014;105:20—33.

[146] Coates RW, Delwiche MJ, Broad A, Holler M. Wireless sensor network with irrigation valve control. Comput Electron Agric 2013;96:13—22.

[147] Guan Y, Chen D, He K, Liu Y, Li L. Review on research and application of variable rate spray in agriculture. In: 2015 IEEE 10th conference on industrial electronics and applications. IEEE; 2015. p. 1575—80.

[148] Heisel T, Andreasen C, Ersbøll AK. Annual weed distributions can be mapped with kriging. Weed Res 1996;36:325—37.

[149] Rew LJ, Cussans GW, Mugglestone MA, Miller PCH. A technique for mapping the spatial distribution of Elymus repots, with estimates of the potential reduction in herbicide usage from patch spraying. Weed Res 1996;36:283—92.

[150] Lan Y, Shengde C, Fritz BK. Current status and future trends of precision agricultural aviation technologies. Int J Agric Biol Eng 2017;10:1—17.

[151] Castaldi F, Pelosi F, Pascucci S, Casa R. Assessing the potential of images from unmanned aerial vehicles (UAV) to support herbicide patch spraying in maize. Precis Agric 2017;18:76–94.

[152] Huang H, Deng J, Lan Y, Yang A, Deng X, Wen S, et al. Accurate weed mapping and prescription map generation based on fully convolutional networks using UAV imagery. Sensors 2018;18. https://doi.org/10.3390/s18103299.

[153] Zheng Y-Y, Kong J-L, Jin X-B, Wang X-Y, Su T-L, Zuo M. CropDeep: the crop vision dataset for deep-learning-based classification and detection in precision agriculture. Sensors 2019;19:1058.

[154] Cap QH, Tani H, Kagiwada S, Uga H, Iyatomi H. LASSR: effective super-resolution method for plant disease diagnosis. Comput Electron Agric 2021;187:106271.

[155] Olsen A, Konovalov DA, Philippa B, Ridd P, Wood JC, Johns J, et al. DeepWeeds: a multiclass weed species image dataset for deep learning. Sci Rep 2019;9:1–12.

[156] Champ J, Mora-Fallas A, Goëau H, Mata-Montero E, Bonnet P, Joly A. Instance segmentation for the fine detection of crop and weed plants by precision agricultural robots. Appl Plant Sci 2020;8:e11373.

[157] Lameski P, Zdravevski E, Trajkovik V, Kulakov A. Weed detection dataset with RGB images taken under variable light conditions. In: International conference on ICT innovations. Springer; 2017. p. 112–9.

[158] Al-Qurran R, Al-Ayyoub M, Shatnawi A. Plant classification in the wild: a transfer learning approach. In: 2018 International Arab conference on information technology. IEEE; 2018. p. 1–5.

[159] Giselsson TM, Jørgensen RN, Jensen PK, Dyrmann M, Midtiby HS. A public image database for benchmark of plant seedling classification algorithms. 2017. ArXiv Prepr ArXiv171105458.

[160] Espejo-Garcia B, Mylonas N, Athanasakos L, Fountas S, Vasilakoglou I. Towards weeds identification assistance through transfer learning. Comput Electron Agric 2020;171:105306.

[161] Yu J, Sharpe SM, Schumann AW, Boyd NS. Detection of broadleaf weeds growing in turfgrass with convolutional neural networks. Pest Manag Sci 2019;75:2211–8. https://doi.org/10.1002/ps.5349.

[162] Schumann A, Waldo L, Mungofa P, Oswalt C. Computer tools for diagnosing citrus leaf symptoms (Part 2): smartphone apps for expert diagnosis of citrus leaf symptoms. Environ Data Inf Serv 2020;2020:1–2. https://doi.org/10.32473/edis-ss691-2020.

[163] Alam MS, Alam M, Tufail M, Khan MU, Güneş A, Salah B, et al. TobSet: a new tobacco crop and weeds image dataset and its utilization for vision-based spraying by agricultural robots. Appl Sci 2022;12:1308.

[164] Liu J, Abbas I, Noor RS. Development of deep learning-based variable rate agrochemical spraying system for targeted weeds control in strawberry crop. Agronomy 2021;11:1480.

[165] Partel V, Costa L, Ampatzidis Y. Smart tree crop sprayer utilizing sensor fusion and artificial intelligence. Comput Electron Agric 2021;191:106556.

CHAPTER 10

Artificial neural modeling for precision agricultural water management practices

Hassan Afzaal[1], Aitazaz A. Farooque[1,2], Travis J. Esau[3], Arnold W. Schumann[4], Qamar U. Zaman[3], Farhat Abbas[5], Melanie Bos[1]

[1]Faculty of Sustainable Design Engineering, University of Prince Edward Island, Charlottetown, PE, Canada; [2]School of Climate Change and Adaptation, Faculty of Science, University of Prince Edward Island, Charlottetown, PE, Canada; [3]Department of Engineering, Faculty of Agriculture, Dalhousie University, Truro, NS, Canada; [4]Citrus Research and Education Center, Institute of Food and Agricultural Sciences, University of Florida, Gainesville, FL, United States; [5]College of Engineering Technology, University of Doha for Science and Technology, Doha, Qatar

10.1 Introduction

Artificial intelligence (AI) has benefited almost all fields of research due to the recent advances in neural computing and numerical modeling. Machines with AI capabilities in a broader scope can perform tasks mimicking human abilities. Systems with AI capabilities can parse data, take actions, and use experiential learning for future decisions. In the development of AI, several branches have emerged, including machine learning (ML), deep learning (DL), and machine vision. ML is a subset of AI, deals with the development of algorithms capable of learning from data without being explicitly programmed. Multilayer perceptron (MLP) is the basic example of an ML algorithm, consisting of an input, hidden and output layer capable of solving basic modeling relationships. Several hydrological studies employed MLP by modeling several regressions and sequence-based problems. Several hydro logical parameters such as surface runoff, river flow, evapotranspiration, and groundwater levels (GWL) exhibit trends along with seasonality treated as a sequence-based problem. MLP and basic ML algorithms handle the nonlinear behavior sequences better than statistical algorithms; however, most of these algorithms do not explain trends and time dependence. The simple architecture of MLP is not capable of storing previous information due to the absence of a memory block. To solve this problem, recurrent neural networks (RNNs) were introduced, capable to store the dynamics of sequences with the help of cycles and node networks [1]. In RNNs, the time-dependence inputs are linked with feedback connections for storing previous information. RNN is capable of storing information of sequence-based problems for short-term data. However, for long sequences and time-series problems, the learning of RNN hinders due to short-term memory problems, also known as the vanishing gradient problem. This problem occurs due to storing all the previous information and the absence of a mechanism to discard the information which is not very useful in model learning. The long-short-term memory (LSTM) neural networks were developed to counter short-term memory and vanishing gradient problems to make it efficient for long sequences. Hydrological studies employed LSTM successfully in catchment areas, runoff modeling, GWL monitoring, and several weather-related studies.

Over time, ML has received tremendous breakthroughs in the last two decades with the invention of technological developments such as graphical processing unit-powered processors, parallel computing, and large-scale sensing devices. Later, a more advanced generation of algorithms, also known as DL, came into existence capable of solving complex image processing problems accurately. DL algorithms are complex to handle, which created opportunities for companies and developers to design DL libraries and frameworks to utilize this domain of AI for the common public. The common DL architecture includes TensorFlow from Google [2], PyTorch from Facebook [3], and MatLab. All these frameworks provide easy implementation of the DL framework for academia, researchers, industry, and common public use. Several open-source platforms for DL provide an opportunity to solve several hydrology-related problems. DL and machine vision algorithms such as convolution neural network (CNN) may be efficiently used in vison-related hydrological problems such as water bodies detection from satellites, land-use land-cover mapping, watershed delineation, urban flood mapping, flood management, and water management resource studies. However, DL-based CNN requires heavy computational resources and a large amount of data for the successful training and implementation of these models. To overcome these issues, reinforcement learning techniques may be utilized as they require fewer data for training and may successfully be implemented in hydrological studies.

In this book chapter, a guide is provided for implementing AI algorithms in the field of hydrology. The introduction section provides the background of AI, ML, and DL and their potential use in various hydrological applications. Section 10.2 provides the classification of different types of a problem encountered in hydrological studies. Section 10.3 introduces various DL frameworks, followed by Section 10.4 providing a gentle introduction to commonly used ML algorithms. Section 10.5 provides a stepwise procedure to implement the ML algorithm from training to implementation. In Section 10.6, a detailed review of the ML algorithm is provided in various hydrological applications. In Section 10.7, ethical concerns and challenges to ML in hydrological studies are provided. Section 10.8 presents a case study from Afzaal et al. [4] to summarize the use of ANNs and DL-based groundwater estimation from major physical hydrology components of two watersheds of Prince Edward Island, Canada.

10.2 Common machine learning problems

In ML, regression analysis predicts the continuous values from one or more predictor variables. Regression analysis draws a relationship between response variables and predictors to explain the continuous fluctuations. Commonly used regression models include linear, polynomial, decision trees, and random forest. Regression was successfully implemented in various applications [5–7] with linear relationships between dependent and response variables.

Classification algorithms can classify the object into distinct categories. Classification develops a mapping function between one or multiple predictors and categorical response variables. Example of classification problems includes spam/nonspam emails, the presence of an object in an image, land-use classification, etc. Naïve Bayes, decision trees, and k-nearest neighbours are the commonly used classification algorithms. Classification-based algorithms may be used in water-related applications such as the classification of water bodies in satellite images, land-use classification, and water resource management. In sequential modeling, an order is imposed on observation to predict one or multiple values along the sequence. Sequences differentiate from the regression problem as in regression problems the order of observation is not important. On the contrary, perseverance of order in the sequence needs to be maintained during training, validation, and testing of sequential ML models. In hydrology, a sequence is the most common ML problem. For example, groundwater, evapotranspiration, moisture, and several environmental parameters are sequence and time series—based problems. Object detection includes the classification and location of the object in the image. Detection problems are complex problems as they involve a higher degree task of object localization in a digital image. Object detection algorithms create a bounding box around the desired object in an image after identifying the presence of an object. Object detection can be implemented in satellite imagery for water management-related tasks. Reinforcement learning is a unique area of ML in which the algorithm learns from the reward and penalty-based policy. Reinforcement learning is bound to experiential-based learning in the absence of training data. Reinforcement learning can be used in several hydrology applications, such as flood prediction, opening, and closing of dams.

10.3 Common deep learning frameworks

Usually, ML algorithms require specified data sequences, model inputs, the definition of loss functions, and optimization of hyperparameters for training, validation, and implementation in real-life problems. The popularity of ML in real-life problems has created opportunities for the development of easily implementing frameworks for ML algorithms for the data scientist, researchers, and community. In this chapter, ML frameworks referred to the platforms for implementing ML algorithms. These frameworks execute complex computational processes efficiently and facilitate higher accuracies. The most common ML frameworks include TensorFlow, PyTorch, Keras, and MatLab.

10.3.1 TensorFlow

TensorFlow is the popular ML framework developed by the Google Brain team [2]. It is a large-scale ML framework that utilizes dataflow graphs for computation in clusters by using multicore central processing units and graphical processing units. TensorFlow initiates operation on an input in the shape of a multidimensional array called a tensor, and the final output is computed after a series of operations performing on the input tensors. This series of operations is termed a graph framework with several advantages, including multiple CPUs and GPUs processing, preservation of graph for later use, and multiple tensor operations. TensorFlow has been favored by various hydrological modelers by virtue of its high flexibility, multilanguage support, portability, and optimized performance [8]. Lim and Wang [9] compared various machine learning algorithms in reproducing temporal and spatial outputs of hydrological models using TensorFlow library. The advanced capabilities of TensorFlow such as model deployment, model visualization using TensorBoard, and model reproducibility have attracted various researchers to use this library in their hydrological models.

10.3.2 Keras

In general, the ML frameworks are written in low-level programming and require a thorough background in implementing and developing ML algorithms. To overcome these issues, high-level application programming interfaces were developed for easy implementation of an ML algorithm. Keras is a high-level application programming interface developed by Chollet [10] to run low-level ML frameworks such as TensorFlow, Theano, and other popular ML frameworks. It is user-friendly, extendable, and easy to work with the framework. Several buildings of neural networks such as layers, loss functions, hyperparameters, and optimizers, are added in Keras to easily build new ML algorithms. Keras has been implemented in various hydrological studies, such as uncertainty in predictions [11], turbulent heat fluxes [12], runoff simulations [13], and uncertainties in hydrological modeling [14].

10.3.3 PyTorch

PyTorch is another DL framework developed by Paszke et al. [3], with a focus on usability as well as speed. PyTorch is a pythonic programming-style framework with easy debugging, consistent computing, and efficient hardware acceleration. PyTorch is designed on dynamic graph definition, which makes faster debugging for the PyTorch framework. PyTorch is famous for research purposes, as it provides several features for experimentation tasks. For example, Li et al. [15] presented a probabilistic neural network for modeling residual errors using the PyTorch library. Other various hydrological studies also used the PyTorch library for various applications [16,17].

10.3.4 MatLab

MatLab is another technical computing framework that is also well-known for vector formulation and matrix problems. It integrates programming, interpretation, computing, and visualization in one platform and can be used for computation, engineering graphics, algorithm development, and application development with graphical user interfaces. Several ML libraries and algorithms may integrate with MatLab for implementing AI for real-life problems.

10.4 Popular machine learning architectures
10.4.1 Multiplayer perceptron

Mulilayer perceptron is a biologically inspired, simplest kind of artificial neural computation model. It consists of several layers, such as input, hidden, and output layers. The purpose of the input layer is to feed the input data into computation or hidden layers. All the processing and modeling take place in hidden layers, followed by an output layer, where the predictions are made for a given problem. All layers in MLP are interconnected with biologically inspired neurons facilitating the computations and predictions. All the biases and weights of MLP are handled through activation functions such as sigmoid, tanh, and Relu. Activation functions are used to boost the values of the input, also known as the firing of a neuron. In the development of MLP, several designing factors such as the number of inputs, the number of hidden layers, the number of neurons, and the activation function are needed to be decided based on the application. Basic structure of MLP is presented in Fig. 10.1. In hydrology studies, MLP can be implemented for regression, and in some cases, for sequence prediction tasks.

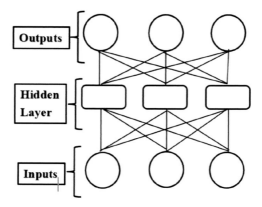

FIGURE 10.1

The basic structure of basic multilayer perceptron, where input, output, and hidden layer relate to neuron connection.

10.4.2 Recurrent neural networks

RNN can store and relate past information and may be efficiently used for sequence or series data. Most of the hydrological problems are sequential based and have some dependence on past data making them fit for the RNN. Parameters sharing, another leverage possessed by RNN, limits the necessity for each node to have its weight. However, for long sequences, RNN faces the vanishing gradient problem hindering the learning curve of the model. To overcome these issues, LSTM neural networks were introduced for long sequences and proved to be a very successful in sequence prediction and forecasting problems. LSTM addresses the vanishing gradient problem by adding more states in their memory blocks. A typical unit of LSTM consists of several states, for example, forget (ft), hidden (ht), and cell state (Ct) to store the sequential dependence of the past data. ft is an important state to keep useful information with the help of activation functions. The detailed structure of the LSTM memory block is presented in Fig. 10.2.

10.4.3 Convolutional neural networks

CNN is very successful and well-known for classification and matrix prediction tasks. It consists of several layers, including convolutional, max pooling, and fully connected layers. CNN can further categorize into classification, detection, and segmentation problems. In classification, usually, the presence of an object is determined with some probability measure. In image detection and segmentation, the object location and masks are identified in an image for advanced processing.

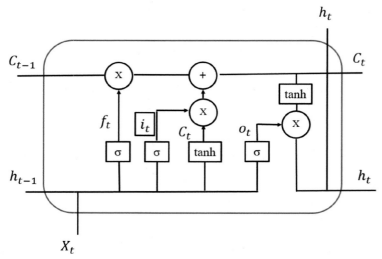

FIGURE 10.2

The basic structure of long short-term memory neural network, where f_t is forget state, C_t is cell state, and h_t is a hidden state.

Most of the applications of CNN can be found in image-related applications and these networks can be used in hydrology and satellite imagery—based studies. For example, the identification of water bodies in satellite images, flood mapping, and runoff modeling may use the leverage of CNN.

10.4.4 Gradient boosting

Gradient boosting is a powerful predictive modeling technique and proved to be very efficient in several predictive problems. The idea of boosting comes from the need for a weak learner to be modified into a better one. In boosting algorithms, usually precise and accurate prediction rules are combined with inaccurate rules of thumb. Gradient boosting algorithms require a loss function to be optimized, a weak learner for prediction, and an additive model for precise estimation. Adaptive gradient boosting algorithms were the first boosting algorithms that received great success.

10.5 Machine learning model development steps
10.5.1 Data preprocessing

Data is the most important element for ML models and require considerable time and effort to make it available for the algorithms. Data preprocessing mitigates the noise in data ensuring a significant improvement in model performance [18]. Data should be thoroughly investigated for potential outliers, missing values, noise, and variability. Data standardization and normalization are recommended techniques for limiting the ranges of all variables to be handled easily by the activation functions. Minns and Hall [19] recommended the data scaling ranges should be proportional to the limits of the activation function in the output layer. Other noise reduction techniques used in several hydrology applications include wavelet transformation [20] and nonlinear techniques [21].

10.5.2 Feature importance

Feature importance is another important step to filter out the variables affecting model performance. Datasets for ML are quite large and require some screening steps to make the training process effective and computationally less expensive. Several methods have been employed to determine the best-suited input variables including priori knowledge-based methods, correlation-based methods, knowledge extraction, and heuristic approaches [22]. Priori knowledge-based methods have been widely applied widely in hydrological studies [18]. The feature importance of priori knowledge—based methods is contingent on the modeller knowledge and case-dependent approach [23]. Correlation-based feature importance methods are also widely used, where there is insufficient information available between dependent and response variables. Correlation methods may further be categorized into cross-correlation [24], partial autocorrelation [25], autocorrelation [26], and nonlinear correlation methods, including mutual information [27] and partial mutual information [28]. In heuristic approaches, the models are trained with different input combinations until maximum accuracy is achieved. In the stepwise variable selection approach, the models are trained for each individual variable [29]. The stepwise approach may further be categorized into forward and backward selection approaches. Most of the heuristic methods are based on hit and trial, and are computationally expensive, limiting the global approach for hydrological models. Knowledge extraction methods are also commonly used feature importance methods used for various hydrological modeling applications. For example, the knowledge can be extracted from trained ML models using the sensitivity method [30]. In sensitivity methods, features are selected based on graphical presentation of responsiveness and human assessment [31]. The method on the selection of feature importance varies with the nature of the hydrological application and thus can be used with the combination of above-discussed methods.

10.5.3 Data split and model development

In ML, the dataset is usually split into training, validation, and testing phases. In the training phase, the ML algorithms learn to form actual values. Validation dataset differentiates from the test dataset as the former provides an evaluation while tuning model hyperparameters. The test set provides an evaluation of unseen and separate data from training and validation phases. Several data split methods have been utilized in ML such as the "hold-out method" and "cross-validation." The hold-out method involves withholding part of the data for validation. When significant convergence is observed in the validation set, a different subset is withheld until the network attains the generalization for the whole dataset. In cross-validation technique, the dataset is split into train, validation, and test set. To prevent overfitting, the test set is used in various configurations and can halt the training process with optimal network architecture. Cross-validation method is suitable when a large amount of data is available.

Designing an optimal ML model is the most challenging task. Typically, the ML model consisted of an input layer, hidden layer, and output layers with several neurons interconnected with all the layers. Determination of an optimal number of hidden layers, several neurons, and other designing factors need to be decided. The optimal structure of neural networks may be determined using global and stepwise methods. In a stepwise method, a trial-and-error approach is adopted to determine the best-suited network architecture. In this method, a basic network is adopted to run multiple trials with different combinations of hidden layers and the number of neurons. Pruning is a stepwise method that usually starts with the simplest structure and proceeds with more complex settings unless there is no improvement observed by employing more combinations. Pruning is a computationally expensive method and requires high computational power. In global methods, knowledge-based principles are implemented in network design. Particle swarm and differential evaluation are examples of global methods.

10.5.4 Hyperparameter tuning

Usually, ML algorithms require an optimal set of model parameters also known as a hyperparameter, to map the relationships between input and output. Several hyperparameters are important to consider, such as random seed, weight initialization, optimizers, and loss function. Random seed should start with a fixed number to make reproducible results. Usually, several ML libraries involve several factors of randomness. By fixing the random seed initialization, similar results may achieve by running the algorithm multiple times, which is an absolute necessity for scientific studies. Weight initialization is another factor to consider while training neural networks. Two well-known methods may be employed in the weight initialization of neural networks, namely transfer learning and training from scratch. In training from scratch methods, neural networks start learning input-output relationships without any prior knowledge. This method is computationally expensive and requires a lot of time and energy during training. Alternatively, the transfer learning approach may be used in the weight initialization method by transferring the knowledge from previously learned objects. This technique is a useful, quick, and computationally less-expensive technique. In training neural networks, optimization algorithms or optimizers are significant. Several well-known optimization algorithms include backpropagation, Adam, Levenberg Marquardt, and conjugate algorithms. The loss function is another parameter to consider during training neural networks as it gives the estimate of deviation between actual and predicted value. Fig. 10.3 shows the stepwise procedure of neural network training.

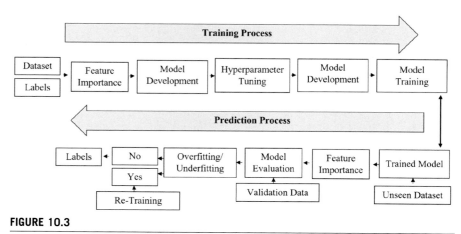

FIGURE 10.3

Stepwise procedure to be employed during neural network training.

10.6 Application of machine learning in different hydrological fields

10.6.1 Rainfall-runoff modeling

Rainfall-runoff modeling is one of the major tasks in the field of hydrology. The trend of ML is increased in rainfall-runoff modeling with the development of advanced ML algorithms and frameworks. Rainfall-runoff modeling is categorized into a sequence-based problem as the quantitative values of surface water fluctuate concerning time. RNN and LSTM are the most common algorithms used for time-series prediction or forecasting tasks. For example, Nguyen and Bae [32] proposed a framework of a coupling forecasting system and developed a hydrological model to predict water levels using LSTM in the urban catchment area. A database consisting of 24 heavy rainfall events was used to train and test the LSTM model. The results indicated the possibility of 3 hours of mean areal precipitation forecasts and flood predictions. Kumar et al. [33] tested LSTM for monthly rainfall forecasting using long sequence data. They used the monthly average precipitation data for years 1871−2016 to train and test the models on different homogenous regions of India. The results suggested the applicability of LSTM in the field of hydrology and associated fields to mitigate climate change risks. Qin et al. [8] developed an MLP and LSTM model to predict the hydrological time series. The performance of the MLP model remained better than the LSTM model. Wang et al. [34] tested theory-guided neural networks for the estimation of subsurface flow. The matrix-based supervised neural network was trained with available observation based on governing equations, engineering controls, and expert knowledge. The trained tested theory-guided neural network performed better than other deep neural networks as the former provided readily generalized predictions. In this study, several two-dimensional saturated flow cases were introduced and tested on the trained neural networks. The results depicted the feasibility of theory-guided neural networks in subsurface flow prediction with heterogeneous model parameters. In another study by Zhu et al. [35], a probabilistic LSTM coupled with a Gaussian process was proposed for daily streamflow forecasting. The proposed method consisted of the biases from the LSTM and retained the probabilistic properties from Gaussian processes. For comparisons, linear models, heteroscedastic Gaussian processes, and vanilla LSTM were developed and tested. The accuracy of probabilistic LSTM was improved, which is of great significance to water resources management and planning.

10.6.2 Groundwater level modeling

GWL helps water resource engineers, policymakers, and managers in decision-making by providing useful information about the aquifer. Modeling of GWL is a complex task that involved thorough knowledge of hydrological components, data sources, modeling procedure, associated inputs, and the physical structure of watershed [36]. Several physical and empirical procedures have been adopted in the literature for GWL estimation. Mohammadi [37] compared the artificial neural network (ANN) and numerical-based model MODFLOW to estimate monthly GWL. The results recommended the better applicability of ANNs in GWL modeling with few input data and time. Several studies in GWL modeling are evident of a better alternative to ANN over physical models [38]. Another study by Mohanty et al. [39] found that neural networks are better for short-term GWL predictions than physical-based models. Karandish and Simunek [40] tested AI-based methods with physical modeling and recommended AI over traditional physical models, namely HYDRUS-2D. Various AI-based methods have been tested by many researchers because of their ease of use and satisfied performance in different geographical regions.

RNN was designed for sequential and time-series data consisting of a memory block for storing useful information. RNN was successfully implemented in GWL studies. For example, Coulibaly et al. [41] tested three different types of neural networks for monthly GWL prediction using temperature and precipitation as input variables in the Gondo aquifer. Among radial basis function networks, generalized radial basis functions, and probabilistic neural networks, RNNs performed efficiently in GWL simulations. Zhang et al. [42] compared basic ANN with LSTM to simulate GWL in China. Temperature, evaporation, water diversion, and time were used as input data to train and validate the neural networks. They used monthly water diversion, evaporation, precipitation, temperature, and time as input data to predict water table depth. The LSTM achieved higher accuracies in predicting GWL in comparison with other algorithms.

Although RNN is specifically designed for sequence-based time-series tasks. However, some studies suggested the better performance of basic ANN over advanced RNN. For example, Muller et al. [43] compared basic ANN with RNN to predict GWL in California, USA. Three different model optimization methods were tested to include random sampling method and the surrogated-based optimization method. Time series of temperature, precipitation, and streamflow were used as input to ML models to estimate fluctuations in GWL with varying time. The results suggested the better results of basic ANN over other compared methods in this study. In another study by Babu et al. [44] comparison of three different algorithms was made to forecast GWL in India. The tested algorithm includes Levenberg—Marquardt, adaptive learning, and gradient descent method. The performance of Levenberg—Marquardt was better than the other tested algorithms.

The CNN is also a popular neural network successfully implemented in hydrological studies. CNN consists of a high number of hidden layers with complex operations such as convolution, max pooling, and batch normalization layers. There is relatively limited literature available on GWL simulation with CNN in comparison with ANN and LSTM. Lahivaara et al. [45] tested one-dimensional CNN to estimate the groundwater storage and GWL using seismic data. Galerkin method was used for wave propagation, followed by feature estimation using CNN. The results suggested the ability of CNN to extract additional information related to features used in this study. In another study by Afzaal et al. [4], CNN was tested along with basic ANN and LSTM to predict GWL in Prince Edward Island using temperature, precipitation, stream levels, and streamflow as input data. The results suggested the better performance of basic ANN over advanced algorithms.

10.6.3 Evapotranspiration modeling

Reference evapotranspiration (ETo) is an important element to consider in water balance and surface energy calculations. Accurate information of ETo is necessary for irrigation planners, growers, water resource managers, and policymakers in irrigation design, water distribution systems, and resource management. Traditionally, lysimeters were the common source to estimate ETo. Nowadays, the use of a lysimeter in ETo estimation is limited due to high installation, maintenance, and operation costs [46]. Various mathematical models are developed to indirectly estimate ETo and are the better alternatives to direct methods due to ease of use [47]. In several research studies, ANNs have been successfully implemented in ETo estimation using climatic variables as input to neural networks. For example, Sudheer et al. [48] estimated the evapotranspiration specifically for rice crops by introducing crop factors in ETo. A radial-based ANN was used to model the ETo using several climatic variables as input. An explicit ANN was proposed by Aytek et al. [49] to simulate the ETo by using the daily climatic variable as an input to the neural network in California, USA. The explicit neural network performed better than the six different conventional ETo estimation methods. Rahimikhoob [50] tested the effect of several climatic variables on ETo estimation using ANNs and compared them with the Penman Monteith method in the region of northern Iran. The result suggested that the air temperature was the most influential variable in explaining the variability in ETo estimation.

ETo is the sequence-based problem and reflects the seasonality and trend in its behavior. Most of the ANNs are unable to consider the time dependence and seasonality factor in their computations. The structure of ANN does not consist of the memory blocks to store the records and extract useful information such as trends or seasonality. RNN proved very effective in time-series problems such as ETo estimation [51]. Afzaal et al. [52] tested LSTM and bidirectional LSTM in computing ETo with various climatic variables in Prince Edward Island. Based on subset regression analysis, maximum temperature, and relative humidity were found to be the most influential variables affecting ETo. The performance of neural networks in this study was found to be satisfactory. In another study by Yin et al. [53], a bidirectional LSTM was tested in short-term ETo forecasting. The models were trained using three variables from three meteorological sites in central China. The results suggested that the bidirectional LSTM trained with sunshine duration, minimum and maximum temperature provides the best forecasts for the selected meteorological sites. In another study by Chen et al. [54], a deep neural network, temporal CNN, and LSTM were tested to estimate daily ETo in the northeast plain in China. The results suggested the higher coefficient of determination recorded for temporal

CNN in comparison with the other tested neural architectures. The studies reflect better performance of ANN over other traditional algorithms and can be successfully implemented in similar hydrological applications.

10.6.4 Water resource management

Management of water supplies in urban areas is a complicated task and requires the use of advanced management tools. Efficient working of wastewater treatment plants, urban water supplies, and conveyance systems is necessary. Zhang et al. [55] proposed a novel approach based on inter-catchment wastewater transfer (ICWT) for sewer overflow mitigation. This study first tested the effectiveness of developed ICWT, followed by the redistribution of inflow in the wastewater treatment plant. For the management of inflow in ICWT, the LSTM is employed to predict the estimated inflow in the ICWT. Another study by Zhang et al. [55] used the internet of things and RNN to estimate the combined sewer overflow in urban areas. The input data for the model was collected from the internet of things combined sewer overflow structure. The neural networks including, MLP, wavelet neural network, LSTM, and gated RNN, had been trained and evaluated in the estimation of combined sewer overflows. Higher accuracies in estimation were achieved using LSTM and gated RNN. Gated RNN resulted in faster learning in the estimation of combined sewer overflow in comparison to LSTM. Fang et al. [56] developed a CNN to forecast the multiple leakage points in the water distribution system. It uses the historical leakage data for training and trained model implemented on real-time data to detect the leakages based on learning features. The trained models presented satisfactory performance on detection of one, two, and three leakage points identification. Nam et al. [57] implemented a reinforcement-based DL technique to develop an autonomous optimal trajectory searching system for bioreactor plants. A deep Q-network was able to optimize the plant efficiency using an optimal trajectory by minimizing the aeration energy. The proposed system was able to decrease the plant energy consumption by 34% reflecting the potential of ML in solving water resource optimization problems. Xu et al. [58] tested an LSTM-based deep neural network to predict the condition of the water supply network. The model input included the pressure and flow measurements at each entry point. To further increase the feature extraction procedure, a parallel LSTM-based network was proposed in this study. In another study by Mamandipoor et al. [59], LSTM-based deep neural network was developed to monitor and identify the faults in wastewater treatment. In this study, modeling faults in the oxidation and nitrification process were identified by capturing the temporal behavior in sensor data. Over five million dataset points were used in this study to achieve the 92% faults detection rate. In water resource management, LSTM and sequence-based algorithms are the most used architecture. The performance of LSTM-based architectures is found to be better than traditional ML applications in most cases.

10.7 Ethical concerns and challenges in machine learning

ML algorithm interpretation is a major challenge in understanding the modeling behaviors in hydrological studies, as ML models are usually termed black-box models. Several visualization libraries have emerged to see the modeling behaviors between several input and response variables. Selection and determination of optimizing hyperparameters for successful ML training and implementation is a difficult and computationally expensive task. Furthermore, there are no specific guidelines available on network development, such as the number of layers, number of neurons, and other design factors. ML and AI models can capture the nonlinear behaviors in data making it efficient in predictive modeling. However, most ML cannot extract the feature automatically and require the inputs to be selected manually. In machine vision problems, DL problems may extract the data automatically. However, high-resolution labeled and supervised data is limited in the water sector, which is the key reason for using the advanced algorithm in water sciences.

In the water sector, the use of AI is modest in comparison with the sectors like energy, healthcare, and transportation [60]. There has been limited work conducted on the responsible use of AI in the water sector. Some of the initial guidelines are provided by Doorn [60] on the responsible use of AI in the water sector. The four most prominent principles of responsible use of AI are transparency, justice, responsibility, and nonmaleficence [60]. The applications of these principles will help the water resource managers, policymakers, and government sectors to lay out the framework for the responsible use of AI in the water sector.

10.8 A case study
10.8.1 Machine learning—based groundwater estimation

A portion of Afzaal et al. [4] is presented as a case study as the activities performed to conduct this study are practical applications of the theory of AI and DL described in this chapter. This study precisely estimated GWL using AI and DL in relatively noncontiguous watersheds. GWL was estimated with DL and ANNs namely MLP, LSTM and CNN, with different variable combinations for Baltic River and Long-Creek watersheds located in Prince Edward Island, Canada.

Different variables such as stream flow, stream level, temperature, precipitation, and evapotranspiration data for seven years (2011–17) were used for training and testing of machine learning models. The hyperparameters were identified using hit-and-trail approach for training (2011–15) and testing (2016–17) periods.

Stream level was the major contributor to fluctuations in GWL among all variables for both watersheds tested in this study; for example, R^2 was recorded as 50.8% and 49.1%, for the Baltic and Long Creek watersheds, respectively. From three AI models, the performance of MLP remained better with reduced RMSE, for example, 0.471 and 1.15 for the Baltic and Long Creek watersheds, respectively.

RMSE was improved by increasing the number of variables from one to four for both watersheds, for example, 1.6% for Long Creek and 11% for Baltic watershed. The results suggested that the stream level was the highly correlated factor in GWL fluctuations.

The results suggested that the GWL can be modeled with stream level efficiently in absence of GWL level data. For the Long Creek watershed, the high RMSE were recorded in comparison with the Baltic watershed. The AI models were unable to predict lower depression in the summer season. The main reason could be because of lurking variables such as pumping data from irrigation wells. The nonavailability of pumping data—this lurking variable was not included in this study.

10.9 Conclusion

In this study, a practical guide is provided to employ ML and AI-based algorithms in water science. This book section covers the basics of ML by providing a stepwise guide for the successful implementation of ML algorithms in various domains of hydrology. The various DL frameworks and libraries provided a breakthrough in implementing AI in hydrology science. The improvement in ML algorithms design provided a broad scope of problems where several hydrology-related problems may resolve. The literature review suggested that the ML algorithm can be implemented in various hydrology problems. However, limited data availability related to hydrology science hinders the application of several advanced DL algorithms. Furthermore, in the water sector, the use of AI is modest in comparison with the sectors like energy, healthcare, and transportation. There is a need to develop guidelines in the responsible use of AI in water-related science.

References

[1] Lipton ZC, Berkowitz J, Elkan C. A critical review of recurrent neural networks for sequence learning. arXiv preprint arXiv:1506.00019; 2015. p. 29.
[2] Abadi M, Barham P, Chen J, Chen Z, Davis A, Dean J, et al. TensorFlow: a system for large-scale machine learning. In: 12th USENIX symposium on operating systems design and implementation (OSDI 16); 2016.
[3] Paszke A, Gross S, Massa F, Lerer A, Bradbury J, Chanan G, et al. Pytorch: an imperative style, high-performance deep learning library. Adv Neural Inf Process Syst 2019; 32.
[4] Afzaal H, Farooque AA, Abbas F, Acharya B, Esau T. Groundwater estimation from major physical hydrology components using artificial neural networks and deep learning. Water 2019;12(1):5.
[5] Rong S, Bao-Wen Z. The research of regression model in machine learning field. MATEC Web Conf 2018;176:01033.
[6] Steyerberg EW, van der Ploeg T, Van Calster B. Risk prediction with machine learning and regression methods. Biom J 2014;56(4):601−6.

[7] Yildiz B, Bilbao JI, Sproul AB. A review and analysis of regression and machine learning models on commercial building electricity load forecasting. Renew Sustain Energy Rev 2017;73:1104−22.

[8] Qin J, Liang J, Chen T, Lei X, Kang A. Simulating and predicting of hydrological time series based on TensorFlow deep learning. Pol J Environ Stud 2019;28(2).

[9] Lim T, Wang K. Comparison of machine learning algorithms for emulation of a gridded hydrological model given spatially explicit inputs. Comput Geosci 2022;159:105025.

[10] Chollet F. Others. Keras [Internet]. GitHub. 2015. Available from: https://github.com/fchollet/keras.

[11] Althoff D, Rodrigues LN, Bazame HC. Uncertainty quantification for hydrological models based on neural networks: the dropout ensemble. Stoch Environ Res Risk Assess 2021;35(5):1051−67.

[12] Bennett A, Nijssen B. Deep learned process parameterizations provide better representations of turbulent heat fluxes in hydrologic models. Water Resour Res 2021;57(5). e2020WR029328.

[13] Fan H, Jiang M, Xu L, Zhu H, Cheng J, Jiang J. Comparison of long short term memory networks and the hydrological model in runoff simulation. Water 2020;12(1):175.

[14] Santos L, Silva E, Freitas C, Bacelar R. Uncertainties propagation in a hydrological empirical model. In: EGU general assembly conference abstracts; 2020. p. 1894.

[15] Li D, Marshall L, Liang Z, Sharma A, Zhou Y. Characterizing distributed hydrological model residual errors using a probabilistic long short-term memory network. J Hydrol 2021;603:126888.

[16] Azmi E, Ehret U, Weijs SV, Ruddell BL, Perdigão RA. "Bit by bit": a practical and general approach for evaluating model computational complexity vs. model performance. Hydrol Earth Syst Sci 2021;25(2):1103−15.

[17] Li D, Marshall L, Liang Z, Sharma A, Zhou Y. Bayesian LSTM with stochastic variational inference for estimating model uncertainty in process-based hydrological models. Water Resour Res 2021;57(9). e2021WR029772.

[18] Oyebode O, Stretch D. Neural network modeling of hydrological systems: a review of implementation techniques. Nat Resour Model 2019;32(1):e12189.

[19] Minns AW, Hall MJ. Modélisation pluie-débit par des réseaux neuroneaux artificiels. Hydrol Sci J 1996;41(3):399−417.

[20] Kışı O, Cımen M. Evapotranspiration modelling using support vector machines. Hydrol Sci J 2009;54(5):918−28.

[21] Sivakumar B, Phoon KK, Liong SY, Liaw CY. A systematic approach to noise reduction in chaotic hydrological time series. J Hydrol 1999;219(3−4):103−35.

[22] Bowden GJ, Dandy GC, Maier HR. Input determination for neural network models in water resources applications. Part 1—background and methodology. J Hydrol 2005; 301(1−4):75−92.

[23] Corominas L, Garrido-Baserba M, Villez K, Olsson G, Cortés U, Poch M. Transforming data into knowledge for improved wastewater treatment operation: a critical review of techniques. Environ Model Software 2018;106:89−103.

[24] Aqil M, Kita I, Yano A, Nishiyama S. A comparative study of artificial neural networks and neuro-fuzzy in continuous modeling of the daily and hourly behaviour of runoff. J Hydrol 2007;337(1−2):22−34.

[25] Yaseen ZM, Fu M, Wang C, Mohtar WH, Deo RC, El-Shafie A. Application of the hybrid artificial neural network coupled with rolling mechanism and grey model

algorithms for streamflow forecasting over multiple time horizons. Water Resour Manag 2018;32(5):1883—99.
[26] Londhe S, Charhate S. Comparison of data-driven modelling techniques for river flow forecasting. Hydrol Sci J 2010;55(7):1163—74.
[27] Bhattacharya B, Solomatine DP. Neural networks and M5 model trees in modelling water level—discharge relationship. Neurocomputing 2005;63:381—96.
[28] Fernando TM, Maier HR, Dandy GC. Selection of input variables for data driven models: an average shifted histogram partial mutual information estimator approach. J Hydrol 2009;367(3—4):165—76.
[29] Pektas AO, Cigizoglu HK. Long-range forecasting of suspended sediment. Hydrol Sci J 2017;62(14):2415—25.
[30] Dibike YB, Coulibaly P. Temporal neural networks for downscaling climate variability and extremes. Neural Network 2006;19(2):135—44.
[31] Cao MS, Pan LX, Gao YF, Novák D, Ding ZC, Lehký D, et al. Neural network ensemble-based parameter sensitivity analysis in civil engineering systems. Neural Comput Appl 2017;28(7):1583—90.
[32] Nguyen DH, Bae DH. Correcting mean areal precipitation forecasts to improve urban flooding predictions by using long short-term memory network. J Hydrol 2020;584: 124710.
[33] Kumar D, Singh A, Samui P, Jha RK. Forecasting monthly precipitation using sequential modelling. Hydrol Sci J 2019;64(6):690—700.
[34] Wang N, Zhang D, Chang H, Li H. Deep learning of subsurface flow via theory-guided neural network. J Hydrol 2020;584:124700.
[35] Zhu S, Luo X, Yuan X, Xu Z. An improved long short-term memory network for streamflow forecasting in the Upper Yangtze River. Stoch Environ Res Risk Assess 2020;34(9):1313—29.
[36] Gao Z, Long D, Tang G, Zeng C, Huang J, Hong Y. Assessing the potential of satellite-based precipitation estimates for flood frequency analysis in ungauged or poorly gauged tributaries of China's Yangtze River basin. J Hydrol 2017;550:478—96.
[37] Mohammadi K. Groundwater table estimation using MODFLOW and artificial neural networks. In: Practical hydro-informatics. Berlin, Heidelberg: Springer; 2009. p. 127—38.
[38] Rajaee T, Ebrahimi H, Nourani V. A review of the artificial intelligence methods in groundwater level modeling. J Hydrol 2019;572:336—51.
[39] Mohanty S, Jha MK, Kumar A, Panda DK. Comparative evaluation of numerical model and artificial neural network for simulating groundwater flow in Kathajodi—Surua Interbasin of Odisha, India. J Hydrol 2013;495:38—51.
[40] Karandish F, Šimůnek J. A comparison of numerical and machine-learning modeling of soil water content with limited input data. J Hydrol 2016;543:892—909.
[41] Coulibaly P, Anctil F, Aravena R, Bobée B. Artificial neural network modeling of water table depth fluctuations. Water Resour Res 2001;37(4):885—96.
[42] Zhang J, Zhu Y, Zhang X, Ye M, Yang J. Developing a long short-term memory (LSTM) based model for predicting water table depth in agricultural areas. J Hydrol 2018;561: 918—29.
[43] Müller J, Park J, Sahu R, Varadharajan C, Arora B, Faybishenko B, et al. Surrogate optimization of deep neural networks for groundwater predictions. J Global Optim 2021; 81(1):203—31.

[44] Maheshwara Babu B, Srinivasa Reddy G, Satishkumar U, Kulkarni P. Simulation of groundwater level using recurrent neural network (RNN) in Raichur District, Karnataka, India. Int J Curr Microbiol Appl Sci 2018;7:3358–67.
[45] Lähivaara T, Malehmir A, Pasanen A, Kärkkäinen L, Huttunen JM, Hesthaven JS. Estimation of groundwater storage from seismic data using deep learning. Geophys Prospect 2019;67(8):2115–26.
[46] López-Urrea R, de Santa Olalla FM, Fabeiro C, Moratalla A. Testing evapotranspiration equations using lysimeter observations in a semiarid climate. Agric Water Manag 2006;85(1–2):15–26.
[47] Abdullah SS, Malek MA, Abdullah NS, Kisi O, Yap KS. Extreme learning machines: a new approach for prediction of reference evapotranspiration. J Hydrol 2015;527:184–95.
[48] Sudheer KP, Gosain AK, Ramasastri KS. Estimating actual evapotranspiration from limited climatic data using neural computing technique. J Irrigat Drain Eng 2003;129(3):214–8.
[49] Aytek A, Guven A, Yuce MI, Aksoy H. An explicit neural network formulation for evapotranspiration. Hydrol Sci J 2008;53(4):893–904.
[50] Rahimikhoob A. Estimation of evapotranspiration based on only air temperature data using artificial neural networks for a subtropical climate in Iran. Theor Appl Climatol 2010;101(1):83–91.
[51] Tawegoum R, Belbrahem R, Chasseriaux G. Modeling evapotranspiration prediction on nursery area using recurrent neural networks. IFAC Proc Vol 2004;37(2):86–91.
[52] Afzaal H, Farooque AA, Abbas F, Acharya B, Esau T. Computation of evapotranspiration with artificial intelligence for precision water resource management. Appl Sci 2020;10(5):1621.
[53] Yin J, Deng Z, Ines AV, Wu J, Rasu E. Forecast of short-term daily reference evapotranspiration under limited meteorological variables using a hybrid bi-directional long short-term memory model (Bi-LSTM). Agric Water Manag 2020;242:106386.
[54] Chen Z, Zhu Z, Jiang H, Sun S. Estimating daily reference evapotranspiration based on limited meteorological data using deep learning and classical machine learning methods. J Hydrol 2020;591:125286.
[55] Zhang D, Hølland ES, Lindholm G, Ratnaweera H. Hydraulic modeling and deep learning based flow forecasting for optimizing inter catchment wastewater transfer. J Hydrol 2018;567:792–802.
[56] Fang Q, Zhang J, Xie C, Yang Y. Detection of multiple leakage points in water distribution networks based on convolutional neural networks. Water Supply 2019;19(8):2231–9.
[57] Nam K, Heo S, Loy-Benitez J, Ifaei P, Yoo C. An autonomous operational trajectory searching system for an economic and environmental membrane bioreactor plant using deep reinforcement learning. Water Sci Technol 2020;81(8):1578–87.
[58] Xu Z, Ying Z, Li Y, He B, Chen Y. Pressure prediction and abnormal working conditions detection of water supply network based on LSTM. Water Supply 2020;20(3):963–74.
[59] Mamandipoor B, Majd M, Sheikhalishahi S, Modena C, Osmani V. Monitoring and detecting faults in wastewater treatment plants using deep learning. Environ Monit Assess 2020;192(2):1–2.
[60] Doorn N. Artificial intelligence in the water domain: opportunities for responsible use. Sci Total Environ 2021;755:142561.

CHAPTER 11

Precision agriculture: making agriculture sustainable

Aneela Afzal[1,2], Mark Bell[3]

[1]*Department of Sociology and Anthropology, PMAS-Arid Agriculture University Rawalpindi, Rawalpindi, Punjab, Pakistan;* [2]*Department of Agricultural Extension, PMAS-Arid Agriculture University Rawalpindi, Rawalpindi, Punjab, Pakistan;* [3]*Strategic Initiatives and Statewide Programs, University of California: Agriculture and Natural Resources, Davis, CA, United States*

11.1 Introduction

Precision Agriculture (PA) is an agricultural management strategy that integrates different strands of knowledge and technology to deliver input combinations that match closely with plant requirements to enable optimal output combinations. Such integration may be facilitated with Decision Support System (DSS)—where a farmer is guided to take a particular set of interventions to address known constraints—to the employment of Artificial Intelligence (AI) and Neural Networks—where automation involving artificial sensors and optimization algorithms that play the role of a virtual farmer in choosing input combinations. So, on one hand, this technology-led management helps decrease input costs, and on the other hand, enhances the probability of greater revenue by increasing yield and matching quality of output to market demand. The advent of modern digital information and communication technologies has enabled PA to become a platform to identify, evaluate, and manage changes within a modern farming system to address multiple management considerations, including sustainability, profitability, and resource conservation.

PA, as a knowledge management system, encompasses an interdisciplinary perspective that seeks to identify, analyze, and manage temporal and spatial variability related to crop production. The primary purpose of PA is to emphasize the benefits of site and time-specific management, and it is not limited to farming per se, and includes policy development, and provision of site-specific advice via modern information and communication technologies. Reduction in environmental degradation is achieved through effective and efficient management of inputs against desired outputs.

PA is also recognizable as satellite farming or site-specific crop management (SSCM). In this setting, it provides an agricultural management approach that involves monitoring, analyzing, and reacting to intra- and intercrop field variability [1]. The challenges of climate change and the pressures to maintain ecological integrity and safeguarding the productive capacity of working landscapes rely on technologies such as conservation tillage, agroforestry, controlled agricultures, etc., as a pathway to minimizing the harmful impact of conventional agriculture. In practical terms, its contribution is in highlighting the importance of variable-rate management interventions in agriculture, where the benefits of fine-tuning management increases substantially when environmental costs are accounted for, implying that the optimality of input use may be more important than previously thought [2].

11.1.1 Evolution of precision agriculture

PA is part of the third wave of the agricultural revolution. Between 1900 and 1930, the first agricultural revolution took place that improved the mechanization system in agriculture. During this period, the focus was to improve farmers' capacity to produce adequate food, to serve the needs of a growing population. The second wave came in the form of Green Revolution that began in the 1960s with conventional breeding and agronomic management. Consequently, farm production rose to levels where an average farmer feeds roughly 155 persons. The Precision is the third wave; the concept of precision farming can therefore be seen as a natural progression in technology-led development that is evolving to meet the challenges of population growth and modernization on one hand and averting natural constraints due to climate change on the other hand. Some argue that the aim is to help raise farm productivity, which enables each farmer to feed about 265 people in the same area [3,4].

PA is evolving at different rates across the world. The United States, Canada, and Australia are considered pioneers of PA, where resource constraints and evolving community preferences of more environmentally senstitive agriculture production is adapting farming systems for sustainable production. The United Kingdom is the first country in Europe to take this step, followed by France, which adopted it in 1997–98. Precision agricultural management strategies have flourished with the invention of GPS and variable-rate spraying systems. Today, most of France's farmers employ variable rate systems in their crops. GPS technologies have facilitated technology solutions with the ability to pinpoint the influence of spatial variability on agricultural productivity and hence creating services that generate specific dose-response mapping to aid site-specific management [2].

In advanced systems, variable rate nutrient application, aerial and satellite imaging, weather prediction, and crop monitoring are part of the initial composite package of resulting PA revolution [5]. The phase of PA combines AI technologies for even more accurate planting, geographical mapping, and soil information. PA gains its influence in areas such as.

- Crop science: by better matching agronomic management to crop demands (e.g., nutrient management),
- Environmental protection: through lowering the environmental hazards and ecological footprint associated with farming (for example, by minimizing nitrogen leaching).

- Economics: through increasing the resource use efficiency through optimal use (e.g., improved nutrient application and other inputs) [6].

Precision agriculture helps farmers to

- Keep track of the resource condition of their farms; make better management decisions; and improve traceability of their products in meeting consumer demand.
- Improve product marketing and profitability.
- Boost the quality of farm produce in meeting specific product attributes (e.g., by altering the protein level in wheat flour in bread and noodle making).

11.1.2 Structure of precision agriculture

In structuring a PA regime, various technologists and farm advisors may utilize a combination of technologies and management information tools such as global positioning systems (GPS), variable rate spraying systems (VRSS), variable rate nutrient application (VRNS), aerial and satellite imaging (SI), weather prediction, and crop monitoring as part of an innovation package. The main components of a PA package (Fig. 11.1) may include the following [1]:

A growing volume of literature suggests that adoption of innovative technologies improves crop yield and input use efficiency thereby reducing environmental degradation.

11.2 Methodology

The work on this chapter was informed by both primary and secondary research. In the primary research, surveys, focus group discussions, and stakeholder interviews were conducted. Secondary research was conducted through a review of literature,

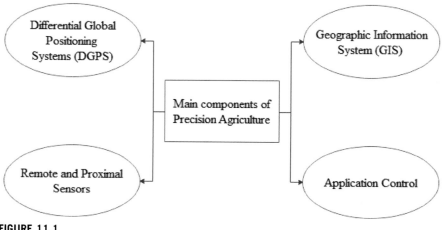

FIGURE 11.1

Main components of precision agriculture (PA) practices.

including journal articles, research reports, and other publications which provided the international, regional, and local context in relation to PA concept and practice. A combination of primary and secondary data was used to assess the financial and economic viability of PA, identify and evaluate impediments to its widespread adoption, discuss measures to overcome these impediments, and to develop a profile of an early adopter of the PA technology, based on sociocultural characteristics.

11.2.1 PA and sustainability objectives of agricultural enterprises

PA technologies could improve the commercial and financial sustainability of agriculture by influencing crop yields in two areas: profitability for farmers and environmental and ecological advantages for the general population.

The founding concepts of PA—the spatial and temporal focus in farm management—provide improved accuracy in agricultural production decisions and a basis for optimizing resource allocations driving agricultural productivity. Zivin et al. [7] describes how PA technologies allow farmers to alter the choice of application and the dosage of fertilizers and other agrochemicals in fields based on temporal and geographical variability of soil attributes. Combined with data on harvested output variability, or yield maps, farmers can conduct economic analyses related to crop output variability in a field under varying climatic conditions to acquire an appropriate risk estimate to guide adaptation decisions to climate variability. For example, a farmer may confirm that 75% of a barley crop grown in a particular field can yield approximately 3.8 tons/ha 70% of the time. Such information can help farmers determine the level of cash return over the expenditures committed for a particular crop in a particular field under a given management regime. By simply incorporating the price of inputs and outputs in an advanced management regime *(technology assisted allocation of inputs to an expected level of output under a planned level of management)* compared to the trial-and-error approach in conventional farming systems, farmers can achieve a greater reliance on farming as a source income, under the guidance of PA. Tang et al. [8], for example, compared variable rate technologies to uniform rate technologies (URT) for phosphorus treatment on soybean and rice in Arkansas and discovered that the profitability of VRT was significantly dependent on both residual phosphorous and soil clay concentration. With such site and management specific information, PA allows greater control of desired outcomes both in the management of farmers' fields as well as in the deployment of environmental regulations, such as in the implementation of regulations for controlled pesticide application [9].

11.2.2 Technology adoption in agriculture

The world's population is projected to reach about 9.2 billion people by 2050, a 34% increase from 2022. To feed the entire population, either the global food production must expand by 70%, or food loss and waste should be minimized across the globe.

The agriculture system, with a narrow focus on production to increase supply, is becoming unsustainable—both financially as well as ecologically. The adoption of PA technology could increase the availability of food without compromising the integrity of the supporting environment and resource stocks [1].

Developing country agricultural systems are usually characterized by small-scale farms with low production and inefficient resource utilization. The agriculture industry of Pakistan has been dealing with several issues in recent years, including decreased crop production, rising input prices, a growing water constraint, power shortages, and limited acceptability of agricultural products in export markets. The environment offers inadequate incentives for prudent use of agricultural input, insufficient information relating to both input and product markets, a lack of quality parameters to guide decisions, and inconsistent agricultural support that distorts resource allocation decisions [10].

Sustainable agricultural systems and market expectations for high-quality goods place the agricultural systems of the developing countries at a point where traditional farming methods alone may not be sufficient to improve farmers' economic situation.

11.2.3 Concept to practice

The private sector is supposed to lead the technology development, commercialization, and adoption. Government is there to facilitate it. As has been the case in developed economies, the private sector may partner with government agencies in developing countries in assisting farmers in PA technology adoption; established farmers will undoubtedly serve as incubators to increase adoption.

In the context of growing demand for quality food, new agricultural technology to improve productivity has a definite advantage [11]. The agricultural system, developed previously with the sole purpose of improving production, is reaching its capacity to provide sustainable improvements in the quality of human life. Gross profits are hiding the costs of agriculture. The adoption of PA technology could prove to be a pathway to rural prosperity which is essential to meet the needs of future generations [12,13].

11.3 Case study of PA adoption in Punjab, Pakistan

Our research study in the Jhelum district of Punjab, Pakistan, witnessed some stark contrasts leading to many interesting insights. Sample data represents that farmers were sowing wheat, maize, and rice as major cash crops in the study area. Fig. 11.2 shows that most farmers, 80.5%, adopted at least one PA practice—represented as adopters, while 19.5% of the farmers did not adopt any PA measure, they are treated as nonadopters.

The major PA farming techniques are "laser land leveling," "water-smart," "zone management," and "soil and nutrient smart." Fig. 11.3 reveals that laser land leveling

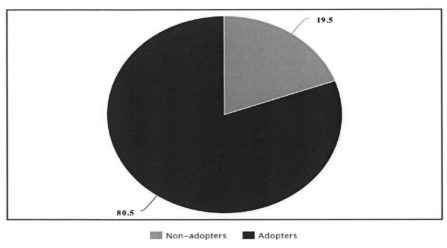

FIGURE 11.2

Adopters and nonadopters of PA practices in the study area of district Jhelum (authors' calculation).

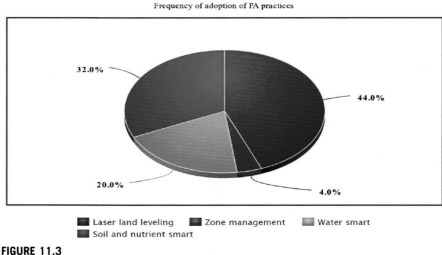

FIGURE 11.3

Frequency of adoption of PA practices (authors' calculation).

is the most adopted (44%) PA practice, it is followed by soil and nutrient smart PA practices, which was adopted by 32% farmers. Water-smart practice is adopted by 20% of farmers and the least adopted (4%) practice is "zone management." However, the PA technology employment is still at the infancy stage; the real potential of the technology will be unleashed when PA is automated through Artificial

Intelligence thus bridging the gap between man and machine. At this stage of PA adoption, a lot of variation may be observed among different farmers in resource consumption and crop yields.

Local farmers are adopting PA techniques and practices according to their funding availability, PA commercial viability, attitude toward technology, and PA equipment market availability along with its after-sales service. Farmers used mobile apps developed by government and private entities to monitor crop productivity zones on their farmland. These mobile apps help farmers take satellite images of the crop zones and also check soil conditions; they also help the farmers detect crop stress from environmental shocks. The zone management system boosts crop performance by monitoring crop growth and the health of the plants to increase yield. Farmers also get assistance from the extension department if they face any problems, though capacity of extension services providers in imparting PA-related training is relatively limited. Some banks including Habib Bank Limited (HBL) and Zarai Tarqiati Bank Limited (ZTBL) are working collaboratively with their creditors in equipping them with PA technology with full-scale service. Large farmers were regularly scouting their fields using drones, while small farmers were restricted to crop scouting apps.

11.3.1 Water-smart

This study shows that 20% of farmers utilize this practice in their farmland. They use information technology, decision support systems and artificial intelligence system for the purpose. Weather forecasting information assists the sample farmers in opening the door to smart irrigation. It helps to utilize water most efficiently and accurately.

11.3.2 Soil and nutrient smart

Fertilizer is a key input to boost plant growth by improving inherent soil fertility. PA practices can help align plant needs to available soil nutrients and use inputs in a balanced and prudent way. Study data shows that targeted fertilizer application is the second most adopted practice by local farmers. Farmers visit the nearest extension centers to collect their soil data to identify the nutrient level in the soil and to know how and when fertilizer should be used. This method is cheaper and more viable for small farmers, especially to increase crop yield and maintain soil health. Data shows that weed control, pests, and disease management are the key challenges for small landholder farmers to protect their crops. This study suggested that small farmers use low-cost and locally available small machine-based variable rate technology at the farm level. Farmers in the study area monitor and adapt practices to address crop variability both within and between fields. It's all about precisely controlling the field conditions to produce more food with fewer resources and lower production costs.

11.4 Evaluation of factors that hamper technology adoption in agriculture

The adoption of technology plays a crucial role in improving sustainability in agriculture. However, research indicates that PA adoption rates are poor. Farmers' decisions to adopt new technology are complex. The complexities of PA adoption procedures may be due to a variety of factors and interactions. Farmers are attracted to PA developments but are less confident of their usefulness, even though they feel PA technologies are beneficial to farming. Technical issues with instruments, lack of suitability of appliances with existing farming activities, lack of access to service software, concerns about service providers misusing agricultural data, challenges in managing the quantity of PA data, user-friendly design features, and cost were all obstacles to PA adoption [14]. Farmers change their mindset about traditional farming practices to adapt to PA technology. Much of the literature is available that provides information with evidence about the adoption of PA in developed agricultural societies, namely North America, Europe, and Australia. Limited studies are available to determine how PA technologies are adopted in developing agricultural societies. Table 11.1 summarizes research that explores the factors affecting landowners and farmers' adoption of PA technologies and highlights the internal, external factors [32].

Table 11.1 Summarization of the factors influencing the farmers for adoption of PA technologies.

S. No.	Factor influencing the farmers for adoption of PA technologies	Description	References
Internal factors			
1	Age	Old-age farmers are more conservative, less eager to adapt, and have shorter planning spans than young farmers.	[15]
2	Gender	Males are more likely to adopt new technologies. Female farmers are frequently left to handle the fields in developing countries when male farmers pursue alternate work, which might exacerbate low levels of adoption.	[16]
	Education	Farmers with a higher level of education can easily be convinced regarding the adoption of PA technologies.	[17]
	Farm size	As larger farm owners can bear expenses, tolerate risks and have higher technological sophistication they are more likely to adopt the PA technologies. Moreover, Some PA technologies such as remote sensing are not appropriate for small farms.	[18]

11.4 Evaluation of factors that hamper technology adoption in agriculture

Table 11.1 Summarization of the factors influencing the farmers for adoption of PA technologies.—cont'd

S. No.	Factor influencing the farmers for adoption of PA technologies	Description	References
	Experience and skills	Farmers require training for acquiring new skills to use novel PA technology like using information and communication systems and interpretation of data outputs. In rural locations, trained and competent agricultural laborers may be limited.	[19,20]
	Farmer needs and values	Farmers' traits and values and product suitability to their specific needs is a significant predictor of adoption.	[21]
	Benefit and risk perceptions	The advantages of PA technologies may be difficult to measure by the farmer. Traditional practices are considered less risky than new technical innovations.	[22]
	Land ownership	Land ownership improves farmer security and encourages a willingness to invest in innovative technologies and management practices.	[23]
	Resources and approach to credit	Easy and cheap credit can encourage the farmer to adopt new technologies.	[17]
External factors			
	Participation in external groups like cooperatives	Farmer participation in external organizations can encourage trial and acceptance of PA technologies.	[19,24]
	Regulations	Facilitating regulations help to speed up the adoption process.	[25]
	Support	The level of assistance provided from the time of acquisition until the time of implementation has a major impact on technology adoption and usage.	[13]
	Trial ability and observability	Evidence suggests that the observation of technology is a key motivation for adoption.	[26]
	Knowledge and information exchange	Farmers who had learned about complex innovations were more willing to implement them. At different phases of the adoption and implementation, different sources of information are crucial, for example, in the awareness and communication phase, mass media is vital, but at later stages technical expertise provided by service providers is significant.	[27,28]

Continued

Table 11.1 Summarization of the factors influencing the farmers for adoption of PA technologies.—cont'd

S. No.	Factor influencing the farmers for adoption of PA technologies	Description	References
	Technology training and task-fit	PA technologies are mostly computer-based applications, which need skill or at least a basic understanding of the computer.	[29]
	Compatibility with current systems and production processes	Compatibility with current technology is critical, especially for modern PA systems. Adoption is also hampered by the incompatibility of hardware and software from various PA vendors.	[28]
	Expenses like financial investments	The initial expenses of PA technologies can be higher than traditional techniques. The expenses are obvious; the financial advantages of PA technology for landowners or farmers can be hard to measure. Moreover, farmers have generally limited financial literacy so doing cost-benefit analysis on behalf of the farmers and communicating it to them in a simple way is crucial.	[30,31]

Adoption of PA technologies is also dependent on the extent of behavioral change necessary for the adopter, that is, the degree to which a novel technique is "destructive" demanding a significant transformation in their current behavior, or "continuously" involving gradual behavior change or offering incentives that complement traditional farming practice. Those technologies are easier to incorporate into existing activities that do not require major behavioral changes [33,34].

In case of emerging economies, government support is essential in ensuring adoption of PA because of the factors such as relatively limited financial resources of farming community, higher financing costs, small farm sizes, unreliable and expensive utilities such as water and power, and limited ecosystem of PA technologies such as dealership networks, after-sales service, and support and financing availability, just to name a few.

11.5 Financial and commercial sustainability of precision agriculture

11.5.1 Evaluation of major cropwise yield gains, cost savings through precision agriculture

The authors analyzed survey data collected from sample farmers in the district of Jhelum. The results show that farmers who adopted at least one or more PA practices had a higher crop yield than farmers who did not. It can be seen in Fig. 11.4.

11.5 Financial and commercial sustainability of precision agriculture

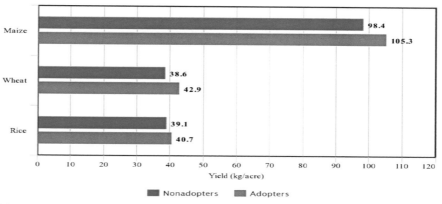

FIGURE 11.4

Adoption of PA practices impact on crop yield (authors' calculation).

The average crop yield increase among the adopters is around 7%. However, we must bear in mind that our dataset is using rudimentary PA technologies such as laser leveling. As the adopters move up in the technology ladder so does their yield. In addition to the crop yield increase, there is also a significant decrease in input consumption thus raising the average profitability to a very reasonable level. Crop yields varied largely, with the minimum wheat yield being 15 kg per acre and the maximum yield is 60 kg per acre, and the maize yield ranges from 70 kg per acre to 130 kg per acre. Rice yields ranged from 30 kg per acre to 55 kg per acre. It is noted that crop yield variation is caused by the farmers' socioeconomic conditions as well as the topographic profile of the region. Overall, adopters received more crop yield in the cases of wheat (11%), maize (7%), and rice (4%). According to the study's findings, farmers who adopted even a single PA practice benefited more than the rest of the farmers. Similar findings were found in the literature.

11.5.2 Evaluation of financial viability of precision agriculture by employing popular capital budgeting techniques

The cost of the agriculture adoptation depends upon the equipment used and the degree of adoption made. According to a study [35], farmers who use precision agriculture techniques for cotton crop production in Brazil increase profits by 3.3% and reduce costs by 6.6% when compared to farmers who do not use PA. Precision methods have resulted in a significant cost reduction (41%) in fertilizer usage. Higher returns and cost reductions demonstrate that PA techniques provide a higher return on investment and economic viability than traditional farming methods. It is also noted that start-up costs in PA equipment applications are quite high, such as sensors, drone spraying, and mapping. Akhtar et al. [36] created a web tool to evaluate the financial viability of farmers who used precision agricultural techniques to increase crop yield while decreasing production costs. The farmers who decided to adopt precision technologies received higher crop yields and income, and their return on investment was higher than the rest of the farmers. However, the assessment criteria make different estimates about the return on investment. It depends on what methodology is used for estimation, such as internal rate of return or net present value. This study suggests that the return on investment for the adopted farmers is higher than the rest of the farmers [37].

11.5.3 Addressing commercial barriers in adoption of precision agriculture

Following the literature review [1,38] and the primary research, the constraints in PA adoption are provided below:

1. Gap between manufacturer and the user, i.e., farmers
2. Lack of training and capacity building.
3. Availability of robust and dynamic data bank.
4. Resource constraint for developing and commercializing new PA tools.
5. Lack of government collaborations to materialize technology transfer.
5. Limited industrial and academic linkages to fulfill the research gap.
6. Nonprovision of services at the doorstep of farmers such as after sale services
7. Cost of the PA equipment.
8. Research programs are not geared towards small landholding farmers, which in certain countries represent a very high percentage of the total farmers.

11.5.4 Environmental impacts of conventional agriculture practices

Agricultural output has a significant impact on ecological processes, both negatively and positively (Table 11.2).

11.5.5 Ecological impacts of conventional agriculture practices

Agriculture has a significant impact on many natural systems. The following are some of the negative consequences of present practices:

Water and wind erosion of exposed topsoil, loss of soil organic matter, diminished biological activity, soil compaction, and a decline in water holding capacity, and salinity of soils in extensively irrigated farmed regions can all contribute to a decrease in soil productivity. Desertification is an increasing concern, especially in regions of Africa, and can be caused by animal overgrazing. Water contaminants from nonpoint sources like salts, fertilizers (particularly phosphate and nitrates), pesticides, and herbicides have been related to agricultural activities. Most of the pesticides have been identified in groundwater, which is specially released from agricultural activities. They're also common in the country's surface waterways.

Table 11.2 Environmental implication of conventional agriculture technologies.

Conventional agriculture technologies	Effect on soil	Effect on water	Effect on biodiversity	Effect on climate or air
Monoculture			Decrease in insect and wildlife habitat, resulting in a greater demand for insecticides. Farmers' capacity to utilize natural pest cycles is harmed, resulting in a greater requirement for pesticides.	
Continuous cropping	Nutrient mining depletes soil fertility.			
Traditional tillage	Increases erosion by reducing soil organic matter.			Degradation of soil organic matter leads to CO_2 emissions.
Intensive Hillside cultivation	Enhances erosion, which leads to degradation of soil.			
Intensive livestock systems	Enhances soil compaction and erosion due to hoof action and overgrazing.	Water quality is degraded by untreated animal manure, and water consumption competes with other demands.	Destroys grassland environment due to overgrazing.	Leads to N_2O and CH_4 emissions due to enteric fermentation.
Inorganic fertilizers	As a result of nitrate leaching, soil acidity intensifies.	Decrease oxygen levels in aquatic habitats owing to runoff; contaminates water for human usage.		Ozone, smog, acid rain, and NO_2 emissions are all caused by this.
Pesticides			By depositing in soils and seeping into water bodies, it endangers animal and human health.	

Continued

Table 11.2 Environmental implication of conventional agriculture technologies.—cont'd

Conventional agriculture technologies	Effect on soil	Effect on water	Effect on biodiversity	Effect on climate or air
Irrigation systems	Waterlogging and salinization caused by insufficient drainage and overirrigation.	Contaminated runoff and excessive water extraction degrade downstream ecosystems.		
New seed varieties	The demand for inputs that have a negative impact on soil may rise.	The demand for inputs that have a detrimental impact on water quality and quantity may rise.	Reduces the ability of landrace types to maintain genetic diversity.	The need for fertilizer may rise, resulting in higher greenhouse gas emissions.
Intensive rice production	Waterlogging, nutritional issues, and salinization are caused by insufficient drainage and floods.	Polluted runoff and excessive water extraction degrade downstream ecosystems.		Because of the anaerobic conditions in paddy fields, it increases CH_4 emissions.
Industrial crop processing		Water demand and the discharge of untreated wastewater impair downstream ecosystems.		The energy needs of machinery contribute to CO_2 emissions.

Many lakes, rivers, and seas are affected by eutrophication and "dead zones" caused by nutrient runoff [39].

Water quality issues influence agricultural output, drinking water supply, and fisheries. Water shortage is caused in many locations by excessive irrigation through groundwater and surface water that can disturb the natural cycle that keeps freshwater availability constant. More than 400 insect and mite pests, as well as more than 70 fungal infections, have developed resistance to one or more insecticides. Pollinators and other helpful insect species have also been harmed by pesticides. This, along with habitat loss because of turning wildlands into agricultural areas, has a significant impact on entire ecosystems (such as the practice of converting tropical rainforests into grasslands for raising cattle). Agriculture's role in climate change has also been studied by many researchers. Excessive carbon dioxide and other greenhouse gases emission can destroy tropical forests and other natural plants. Soils have been discovered to have huge carbon stores in recent investigations.

11.5.6 Environmental impacts of precision agriculture

The following PA technologies play an important role in agriculture expansion without compromising the environment [40,41].

To mitigate erosion concerns, PA technologies are combined with soil erosion controls and digital elevation maps. This system helps to reduce erosion and ultimately protect the environment's salinity, waterlogging, nutrient depletion, flooding, and many other issues.

- The utilization of unmanned aerial vehicles (UAVs) for weed management.
- Thermal and mechanical weed management and herbicide patch application.
- Incorporation of disease and pest control equipment.
- Improved pest management through temporal and spatial interventions; excessive pesticides can leach down to the ground and contaminate the water bodies.
- PA helps manage fertilizers more efficiently and effectively.
- PA enables utilization of historical data on fertilizer, water, and chemical preservation along with crop yields and helps compare it with site-specific crop management.
- PA helps in carefully placing drip irrigation and regulating the system to ensure that water does not come in contact with leafy crops, which prevents contamination through irrigation.
- PA reduces the dangers from microbial contamination or toxin contamination that can damage human health.

11.5.7 Ecological impacts of precision agriculture

"Industrial agriculture" is a word used to describe the agriculture that today pervades the developed world. It is highly automated, frequently monoculture, and reliant on big farms and fields with extensive fertilizer, water, pesticide, and other chemical treatments. Industrial agriculture is not sustainable due to changes in ecosystems. Precision farming could be a better alternative. Precision agriculture maintains ecosystem and makes agriculture more sustainable while also ensuring high water quality, soil biodiversity, soil organic carbon management, and reducing land losses [42].

11.5.7.1 Managing the fertilizer and pesticide applications

PA plays an important role in managing this situation. For example, distant and proximal sensors can now evaluate a crop's nitrogen requirements as well as identify weeds and some agricultural pathogens [43]. Nitrogen and phosphorus, particularly in the form of nitrates, have a significant impact on water quality. The timing and quantity of fertilizer application can be controlled through PA technologies.

- **Soil erosion**

The subsoil is the most fertile layer of the soil, and its loss results in a depleted soil that requires ever-increasing fertilizer inputs to sustain yields. Soil erosion is affected by soil type, drainage, topography, and management. By training farmers on their cultivation techniques, precision agriculture can assist in limiting the impacts of soil erosion.

Organic materials in soil organic matter are essential for sustaining soil structure, water-holding capacity, nutrient status, cation exchange capacity (CEC), and biodiversity in the soil. As biodiversity influences soil processes such as organic matter decomposition, nutrient cycling, hydrology, structural stability, and gaseous exchanges, so biodiversity conservation is critical for sustainable agriculture.

- **Irrigation**

Irrigation is used in many parts of the semi-arid areas. Irrigation draws a lot of the water from aquifers, causing plenty of issues when abstraction is unregulated. This can cause the salinization or sodicity of agricultural land. When excessive irrigation water is supplied to agricultural land and drainage is inadequate, saline groundwater can rise to the surface.

Because of nutrient deficiencies, excess sodium causes dispersion of clay in the soil and the soil structure to disintegrate, resulting in waterlogging and poor crop output. Irrigation requires adequate drainage to overcome these issues. Precision agriculture entails precise irrigation management, to apply water just where and when it is needed.

The automated machines with GPS guidance can potentially reduce soil compaction and carbon foot printing. On hilly terrain, contour cultivation and automatic guidance can aid in the production of vegetation at key locations and crop field borders. It reduces erosion, runoff, surface water, and fertilizers and reduces the risks of flooding. The sensing of manure composting through a variable rate manure application system helps to reduce groundwater pollution. This technology also controls air pollution through the reduction of ammonia emission into the air.

Soil texture mapping for precision irrigation helps to avoid excessive water utilization on-field and waterlogging issues.

- **Herbicide and insecticide application**

Weed mapping or detection helps in locating and spraying herbicides in a specific area. The precise application of herbicides reduces their harmful effects on soil and water bodies. The PA technologies for the detection of diseases in plants, including multisensory optical detection, airborne space detection, and volatile sensors, help in the early detection and treatment of pests or diseases. This technology saves the environment from the excessive application of pesticides and herbicides on fields without any regulation.

The application of tree size and architecture detection sensors on orchard and vineyard precision spraying reduces the application of pesticides by 20%–30%. The precision spray of phosphorous recovers up to 25% of phosphorous [19].

PA practices help farmers increase crop yield by improving crop monitoring, disease, and pest management, and reducing environmental impact. Technology helps to reduce its carbon footprint via land-use management practices. It reduces input resource costs by increasing resource-use efficiency, achieving soil health, and realizing genetic potential in crop production.

11.5.8 The verdict: environmental and ecological impacts of precision agriculture compared with conventional agriculture

Production, soil composition/erosion, biodiversity, water consumption, energy utilization, and greenhouse gas emissions should all be considered when comparing conventional versus precision agriculture. The ecological and environmental protection through adoption of PA is the main concern.

The conventional cropping system, through unregulated application of pesticides and fertilizers, negatively impacts the environment. However, the PA technologies improve the yield through the precise application of pesticides and nutrients while reducing environmental hazards. Biodiversity preservation is another area that needs the serious attention of modern agricultural practices. The conventional agriculture system reduces the biodiversity of pests and microorganisms while precision agriculture discourages the traditional cropping system method and provides sustainable alternatives like intercropping and integrated pest management that promote biodiversity. The preservation of biodiversity stabilizes the ecological cycles.

Globally, water is a renewable resource that can fulfill the demands of our existing population. Locally, however, water is a limited resource that must be used wisely. Although there is a limited amount of fresh water accessible for use across the world, regional dynamics and restrictions make it far more challenging for many millions of people to obtain it. Agriculture consumes over 70% of the water used on the planet. Increased demand for freshwater is putting a strain on world supplies. To protect this resource, PA technologies provide a solution to wisely use the water in the right place at the right time. Better water and soil management through the use of PA technologies may result in much higher crop yields than the conventional farming system.

PA technologies wisely apply the inputs to the crops that reduce the reliance on non-renewable resources for agricultural practices. Conversely, traditional agricultural practices utilize a lot of energy to prepare, produce, and transport food. Energy efficiency is very important to reduce greenhouse gas emissions and lower production costs. It is studied that agricultural activities release 5% of CO_2 and account for 10%–12% of other greenhouse gas emissions. PA technologies can help to reduce these emissions by improving energy efficiency. Today, PA is viewed as an "environmentally friendly system solution that maximizes product quality and quantity while minimizing expense, human involvement, and natural variation" [44].

11.6 Social and economic impact of precision agriculture

11.6.1 Precision agriculture's impact on farm household incomes

Climate change and a growing population are a major challenge for the whole world. The former affects the crop yield but latter demands more food to feed the rising population. To meet these challenges, PA practices are currently adopted in the world to avoid environmental damage and to increase the crop yield. Studies [3] show that the farmers who adopted precision agriculture practices faced higher costs (28%) than the rest of the farmers in San Lorenzo, but their profit was increased by 46%, with net income of 34%. Similarly, Capmourteres et al. [45] studies find that precision technology adoption reduces significant costs of production when farmers adopt it as an adaptation and mitigation option. These practices significantly increase farm income for households as well as the farm economy.

11.6.2 Economic dynamics of adoption of precision agriculture

The study by Tesfaye and Tirivayi [46] combines the crop production data and cost data to analyse the economic dynamics of precision technology adoption at the farm level. This study reveals that farmers who use precision technologies reduce farm risk. As the capital cost these technologies is high, it is more feasible for small farmers to rent them. In Ontario, rent for the precision technology service provider varies from $309 to $741 per ha. Although the price is high due to commodity market conditions, farmers get a benefit when they receive higher returns.

A study [47] shows that some precision technologies have low break-even points due to rising rental and managerial costs, but the findings from the sample farmers show that a net gain of $30 to $50 per ha was achieved by the farmers who adopted PA compared to those who did not. The overall gain from precision agricultural practices depends on the geographical location and ecological zones of the regions. It is noted that the farmers who could not buy PA equipment preferred to take it on a rental basis. The authors calculated rental and equipment costs based on the sample data in the Jhelum district and found that laser land leveling equipment costs vary on average from 600,000 Pakistan rupees to 850,000 Pakistan rupees per piece of equipment. The same is available at an average rental cost of around Pakistan Rupees 2000 per hour to Pakistan Rupee 6000 per hour.

11.6.3 Social and gender dynamics of precision agriculture

Small farmers need to be nudged to adopt PA owing to scarcity of resources. This encouragement may come from government support in the form of lower financing cost of PA equipment, from PA equipment manufacturers through availability of tool and machines on rental basis, or provision on leasing or financing. Importantly, after-sales service and support are essential for any technology to proliferate. So, PA equipment manufacturers will have to develop their dealership networks across the country, develop an ecosystem of PA by training local technicians to be able to provide installation and maintenance services of the equipment and partner with banks to provide easy and cheap financing. Moreover, at the initial stage, strong promotion and effective marketing will definitely help accentuate benefits of the PA equipment in terms of increase in farm efficiency.

Scarcity of financial resources is a major constraint that makes farmers unable to adopt PA [48]. The literature suggests that, in most countries, the agriculture market is dominated by male traders. Therefore, female farmers are unable to take decisions regarding agricultural production [16]. Using survey data from the Jhelum district, the authors calculated differences in social characteristics among the adopters and nonadopters. The data shows that the farmers who were well endowed were able to adopt more than the rest of the farmers. Details of the social characteristics can be seen in Table 11.3.

According to statistics, farmers who adopted at least one or more PA practices have more favorable social attributes than farmers who were unable to adopt them. Those with larger land holdings (16.08), more livestock (3.56), and more years of schooling (8.41) were more willing to adopt than the rest of the farmers (Table 11.3). Land ownership also exhibits interesting findings. The farmers who

Table 11.3 Social characteristics of the sample farmers in the Jhelum district.

Social characteristics	Nonadopters	Adopters
Farm size	9.72	16.08
Livestock ownership	1.11	3.56
Education (years of schooling)	6.55	8.41
Land ownership (ratio of owned land/rented land of the sample farmers)	0.4	2.1

Note: Authors' calculations.

have their own farmland were more intended to adopt than the farmers who have cultivated land on a rental basis. It might be shown that PA practices and technologies are costly in the initial period for farmers. Therefore, comparatively large and progressive farmers who have their own farmland and a large farm size have adopted PA practices. The key motive for adoption is to generate better returns and to achieve economies of scale in case of provision of this technology on rental basis. The social characteristics of the adopted or willing-to-adopt farmers confirm that the government and private institutions should focus more on vulnerable farming households through inclusive policies and practices which can help shift them from nonadopters to adopters.

11.7 Conclusion

In general, the returns in conventional agriculture have been lagging behind industry and services. Thus, it's no surprise that conventional agriculture has been outcompeted by other sectors as a private investments recipient. Our study shows that investment in even rudimentary precision agriculture techniques has a very high return on investment. PA investments have the potential to fetch the returns for agriculture that are comparable to industry and services and thus make it sustainable and profitable in the long run. However, the stakeholders need to be mindful of sociocultural and economic considerations in tapping full potential of the precision agriculture. According to our study, technology affordability is a big hindrance that can be overcome by availability of rental, leasing, and cheaper financing. The affordability is followed by availability of full-scale after-sales services of the technology which may be overcome by building a technology ecosystem, training and capacity building of technicians and farmers. In terms of sociocultural characteristics, the early adopters are generally those farmers who enjoy relatively higher social positions that may be in terms of bigger farm sizes, more livestock holding, higher education, and higher affluence. Vulnerable segments of society such as female and youth are not among the early adopters so stakeholders need to follow inclusive strategies to put this technology in the reach of these segments.

Many factors—such as technical skills, age, and resource constraints—hinder the adoption of precision technologies among farmers. The chapter also investigates the socioeconomic effects of precision technology adoption and its financial impact on farm income. Gender and social dynamics of precision agriculture adoption, constraints, and possible solutions to increase the frequency of adopters were also discussed. This chapter reveals that farmers who have adopted precision technology have received higher crop yields and income, but that it varies with the availability of precision technology at the farm level. Most of the adopters are males because the technology as well as financial capital is concentrated among male farmers; thereby alluding to yet another reason that restricts female farmers from getting optimum production, which ultimately reduces the female farmers' gain per hectare than male-owned farmland. Over the last half century, conventional agriculture practices did improve yields but at the cost of the ecosystem and environmental damage. Therefore, PA provides the most sustainable solution to cope with the environmental challenges and to ensure food security.

References

[1] Bwambale E, Abagale FK, Anornu GK. Smart irrigation monitoring and control strategies for improving water use efficiency in precision agriculture: a review. Agric Water Manag 2022;260:107324.

[2] Lee CL, Strong R, Dooley KE. Analyzing precision agriculture adoption across the globe: a systematic review of scholarship from 1999−2020. Sustainability 2021;13: 10295. https://doi.org/10.3390/SU131810295.

[3] Monzon JP, Calviño PA, Sadras VO, Zubiaurre JB, Andrade FH. Precision agriculture based on crop physiological principles improves whole-farm yield and profit: a case study. Eur J Agron 2018;99:62−71. https://doi.org/10.1016/J.EJA.2018.06.011.

[4] Lal R, Delgado JA, Groffman PM, Millar N, Dell C, Rotz A. Management to mitigate and adapt to climate change. J Soil Water Conserv 2011;66(4):276−85. https://doi.org/10.2489/jswc.66.4.276.

[5] Raza A, Razzaq A, Mehmood SS, Zou X, Zhang X, Lv Y, et al. Impact of climate change on crops adaptation and strategies to tackle its outcome: a review. Plants 2019;8:34. https://doi.org/10.3390/PLANTS8020034.

[6] Bae C, Davis RA, Pipiras V. Periodic dynamic factor models: estimation approaches and applications. Electron J Stat 2018;12(2):4377−411. https://doi.org/10.1214/18-EJS1518.

[7] Zivin G, Cao J, Roberts J, Ying J, Ying Z, Tom-Chang BY, et al. The effect of pollution on worker productivity: evidence from call center workers in China. Archsmith Heyes Saberian 2014;11(1):151−72. https://doi.org/10.1257/app.20160436.

[8] Tang Q, Bennett SJ, Xu Y, Li Y. Agricultural practices and sustainable livelihoods: rural transformation within the Loess Plateau, China. Appl Geogr 2013;41:15−23. https://doi.org/10.1016/J.APGEOG.2013.03.007.

[9] Botta A, Cavallone P, Baglieri L, Colucci G, Tagliavini L, Quaglia G. A review of robots, perception, and tasks in precision agriculture. Appl Mech 2022;3(3):830−54.

[10] Howes MJR, Quave CL, Collemare J, Tatsis EC, Twilley D, Lulekal E, et al. Molecules from nature: reconciling biodiversity conservation and global healthcare imperatives for sustainable use of medicinal plants and fungi. Plants People Planet 2020;2(5):63−481. https://doi.org/10.1002/PPP3.10138.

[11] Sardar A, Kiani KA, Kuslu Y. An assessment of willingness for adoption of climate-smart agriculture (CSA) practices through the farmers' adaptive capacity determinants. Yüzüncü Yıl Üniversitesi Tarım Bilimleri Dergisi. 2019;9(4):781−91. https://doi.org/10.29133/yyutbd.631375.

[12] Mwongera C, Nowak A, Notenbaert AMO, Grey S, Osiemo J, Kinyua I, et al. Climate-smart agricultural value chains: risks and perspectives. Climate-Smart Agric 2019: 235−45. https://doi.org/10.1007/978-3-319-92798-5_20.

[13] Cruz-Garcia GS, Sachet E, Blundo-Canto G, Vanegas M, Quintero M. To what extent have the links between ecosystem services and human well-being been researched in

Africa, Asia, and Latin America? Ecosyst Serv 2017;25:201−12. https://doi.org/10.1016/J.ECOSER.2017.04.005.

[14] Lee WA, Oywaya NA, Mwangi Kibe A, Ngeno Kipkemoi J. Climate-smart agriculture and potato production in Kenya: review of the determinants of practice. Climate Dev 2021;14:75−90. https://doi.org/10.1080/17565529.2021.1885336.

[15] Schleifer P, Sun Y. Reviewing the impact of sustainability certification on food security in developing countries. Global Food Secur 2020;24:100337. https://doi.org/10.1016/J.GFS.2019.100337.

[16] Teychenne M, Apostolopoulos M, Ball K, Olander EK, Opie RS, Rosenbaum S, et al. Key stakeholder perspectives on the development and real-world implementation of a home-based physical activity program for mothers at risk of postnatal depression: a qualitative study. BMC Publ Health 2021;21(1):1−11. https://doi.org/10.1186/S12889-021-10394-8/FIGURES/1.

[17] Aslan MF, Durdu A, Sabanci K, Ropelewska E, Gültekin SS. A comprehensive survey of the recent studies with uav for precision agriculture in open fields and greenhouses. Appl Sci 2022;12(3):1047.

[18] Martinho VJPD, Guiné R de PF. Integrated-smart agriculture: contexts and assumptions for a broader concept. Agronomy 2021;11:1568. https://doi.org/10.3390/AGRONOMY11081568.

[19] Mumtaz Z, Whiteford P. Comparing formal and informal social protection: a case study exploring the usefulness of informal social protection in Pakistan. J Int Comparative Soc Policy 2021:1−30. https://doi.org/10.1017/ICS.2021.9.

[20] Doss C, Meinzen-Dick R, Quisumbing A, Theis S. Women in agriculture: four myths. Global Food Secur 2018;16:69−74. https://doi.org/10.1016/J.GFS.2017.10.001.

[21] Waha K, van Wijk MT, Fritz S, See L, Thornton PK, Wichern J, et al. Agricultural diversification as an important strategy for achieving food security in Africa. Global Change Biol 2018;24(8):3390−400. https://doi.org/10.1111/gcb.14158.

[22] Rosenstock TS, Lamanna C, Chesterman S, Bell P, Arslan A, Richards M, et al. The scientific basis of climate-smart agriculture: a systematic review protocol. In: CCAFS working Paper No. 138. Copenhagen, Denmark: CGIAR research program on climate change, agriculture and food security (CCAFS); 2016. https://cgspace.cgiar.org/handle/10568/70967.

[23] Teklewold H, Mekonnen A, Kohlin G, Di Falco S. Does adoption of multiple climate-smart practices improve farmers' climate resilience? Empirical evidence from the Nile basin of Ethiopia. Clim Change Econ 2017;08(01):1750001. https://doi.org/10.1142/S2010007817500014.

[24] Shafi U, Mumtaz R, Shafaq Z, Zaidi SMH, Kaifi MO, Mahmood Z, et al. Wheat rust disease detection techniques: a technical perspective. J Plant Dis Prot 2022;129:489−504. https://doi.org/10.1007/S41348-022-00575-X/TABLES/3.

[25] Singh R, Singh GS. Traditional agriculture: a climate-smart approach for sustainable food production. Energy Ecol Environ 2017;2(5):296−316. https://doi.org/10.1007/s40974-017-0074-7.

[26] Ojo TO, Baiyegunhi LJS. Determinants of climate change adaptation strategies and its impact on the net farm income of rice farmers in South-West Nigeria. Land Use Pol 2020;95:103946. https://doi.org/10.1016/j.landusepol.2019.04.007.

[27] Monzon LZ, Zhang L, Li W, Zhang J, Frewer LJ. Adoption of combinations of adaptive and mitigatory climate-smart agricultural practices and its impacts on rice yield and

income: empirical evidence from Hubei, China. Clim Risk Manag 2021;32:100314. https://doi.org/10.1016/J.CRM.2021.100314.
[28] Williges K, Mechler R, Bowyer P, Balkovic J. Towards an assessment of adaptive capacity of the European agricultural sector to droughts. Clim Serv 2017;7:47–63. https://doi.org/10.1016/J.CLISER.2016.10.003.
[29] Dell M, Jones BF, Olken BA. Temperature shocks and economic growth: evidence from the last half century. Am Econ J Macroecon 2012;4(3):66–95. https://doi.org/10.1257/mac.4.3.66.
[30] Winters P, Corral L, Gordillo G. Rural livelihood strategies and social capital in Latin America: implications for rural development projects. Working paper series in agricultural and resource economics. Armidale, Australia: University of New England; 2001.
[31] Baio FHR, Da Silva SP, Camolese H, Da S, Neves DC. Financial analysis of the investment in precision agriculture techniques on cotton crop. Eng Agrícola 2017;37(04): 838–47. https://doi.org/10.1590/1809-4430-ENG.AGRIC.V37N4P838-847/2017.
[32] Yin H, Cao Y, Marelli B, Zeng X, Mason AJ, Cao C. Soil sensors and plant wearables for smart and precision agriculture. Adv Mater 2021;33(20):2007764.
[33] Amin A, Nasim W, Mubeen M, Sarwar, Urich P, Ahmad A, et al. Regional climate assessment of precipitation and temperature in Southern Punjab (Pakistan) using SimCLIM climate model for different temporal scales. Theor Appl Climatol 2018; 131(1–2):121–31. https://doi.org/10.1007/s00704-016-1960-1.
[34] Quandt A. Measuring livelihood resilience: the household livelihood resilience approach (HLRA). World Dev 2018;107:253–63. https://doi.org/10.1016/J.WORLDDEV.2018.02.024.
[35] Hashni T, Amudha T, Ramakrishnan S. Smart farming approaches towards sustainable agriculture—a survey. 2022. p. 695–714. https://doi.org/10.1007/978-981-16-7330-6_52.
[36] Akhtar MN, Shaikh AJ, Khan A, Awais H, Bakar EA, Othman AR. Smart sensing with edge computing in precision agriculture for soil assessment and heavy metal monitoring: a review. Agriculture 2021;11:475. https://doi.org/10.3390/AGRICULTURE11060475.
[37] Yost MA, Sudduth KA, Walthall CL, Kitchen NR. Public–private collaboration toward research, education and innovation opportunities in precision agriculture. Precis Agric 2019;20(1):4–18. https://doi.org/10.1007/S11119-018-9583-4/TABLES/4.
[38] Parsons M, Brown C, Nalau J, Fisher K. Assessing adaptive capacity and adaptation: insights from Samoan tourism operators. Clim Dev 2018;10(7):644–63. https://doi.org/10.1080/17565529.2017.1410082.
[39] Lipper L, Thornton P, Campbell BM, Baedeker T, Braimoh A, Bwalya M, et al. Climate-smart agriculture for food security. Nat Clim Change 2014;4:1068–72. https://doi.org/10.1038/nclimate2437.
[40] Padilla-Díaz CM, Rodriguez-Dominguez CM, Hernandez-Santana V, Perez-Martin A, Fernandes RDM, Montero A, et al. Water status, gas exchange and crop performance in a super high density olive orchard under deficit irrigation scheduled from leaf turgor measurements. Agric Water Manag 2018;202:241–52. https://doi.org/10.1016/j.agwat.2018.01.011.
[41] Sardar A, Kiani KA, Kuslu Y, Bilgic A. Examining the role of livelihood diversification as a part of climate-smart agriculture (CSA) strategy. Atatürk Üniversitesi Ziraat Fakültesi Dergisi 2020;51(1):79–87. https://doi.org/10.17097/ataunizfd.604937.

[42] Arslan A, McCarthy N, Lipper L, Asfaw S, Cattaneo A, Kokwe M. Climate smart agriculture? Assessing the adaptation implications in Zambia. J Agric Econ 2015;66(3): 753—80. https://doi.org/10.1111/1477-9552.12107.

[43] Richard B, Qi A, Fitt BDL. Control of crop diseases through integrated crop management to deliver climate-smart farming systems for low- and high-input crop production. Plant Pathol 2022;71(1):187—206. https://doi.org/10.1111/PPA.13493.

[44] Balafoutis A, Beck B, Fountas S, Vangeyte J, Van Der Wal T, Soto I, et al. Precision agriculture technologies positively contributing to GHG emissions mitigation, farm productivity and economics. Sustainability 2017;9(8):1339. https://doi.org/10.3390/SU9081339.

[45] Capmourteres V, Adams J, Berg A, Fraser E, Swanton C, Anand M. Precision conservation meets precision agriculture: a case study from southern Ontario. Agric Syst 2018; 167:176—85. https://doi.org/10.1016/J.AGSY.2018.09.011.

[46] Tesfaye W, Tirivayi N. Crop diversity, household welfare and consumption smoothing under risk: evidence from rural Uganda. World Dev 2020;125:104686. https://doi.org/10.1016/J.WORLDDEV.2019.104686.

[47] Sardar A, Kiani AK, Kuslu Y. Does adoption of climate-smart agriculture (CSA) practices improve farmers' crop income? Assessing the determinants and its impacts in Punjab province, Pakistan. Environm Dev Sustain 2021;23(7):10119—40. https://doi.org/10.1007/S10668-020-01049-6.

[48] Delavarpour N, Koparan C, Nowatzki J, Bajwa S, Sun X. A technical study on UAV characteristics for precision agriculture applications and associated practical challenges. Rem Sens 2021;13(6):1204.

CHAPTER 12

Environment: role of precision agriculture technologies

Shoaib Rashid Saleem[1,2], Jana Levison[1], Zainab Haroon[2]

[1]School of Engineering, University of Guelph, Guelph, ON, Canada; [2]Data Driven Smart Decision Platform, PMAS-Arid Agriculture University Rawalpindi, Rawalpindi, Punjab, Pakistan

12.1 Introduction

Agricultural production rapidly increased after the beginning of green revolution in 1960's [1] due to wide-scale adoption of new technologies, [2] tillage implements, commercial chemical fertilizers, agrochemicals, modern irrigation practices, high yielding grain varieties [2]. The introduction of chemical fertilizers was the game changing factor for many farmers, a fast and friendly replacement for livestock manure, especially in the developing world [3]. In many cases, farmers, started applying these chemical fertilizers without considering the soil and crop requirements and the farmers perception was "apply more fertilizer, get more yield" [4]. Researchers and scientists started to study the impacts of fertilizers and agrochemical spraying on crops and developed optimum windows for nutrient application for many cereal crops and fruit trees [5]. The excessive application has not only reduced the crop productivity but also increased environmental pollution [6,7]. Excessive nutrients can leach down and pollute groundwater, can be transported from fields to pollute surface water, and fertilizer through runoff, and volatilization can result in air pollution [8]. Therefore, it is essential that innovative techniques and modern technologies are adopted to convince farmers and producers to smartly manage their fields.

In the early 1990s, development of global positioning system (GPS) and introduction of fast processing computers provided basis for mapping and visualization of crop areas on large scale [9]. However, GPS usage was limited for military purposes only [10]. Agricultural researchers and scientists started working on soil and crop variability mapping in the early 2000s after the US armed forces allowed public use of GPS technology in other sectors which was the basis of smart or precision agriculture [9]. The development of fast computer processors and artificial intelligence has increased the scope of precision agriculture, to ensure the crop and soil receive exactly what they need for optimum health and productivity. The overall goal of the development of precision agriculture technologies (PATs) was to ensure sustainable, profitable, and environment-friendly agriculture [11]. The reliance upon software, information technology (IT) services, variable rate technologies, and Internet of things (IoT) to process real-time soil data, weather data, and crop yields and variability data are essential in adoption of PATs [12].

Adoption of PATs is also helping farmers to reduce the carbon foot printing and comply with good farming standards [13]. PATs can help to achieve environmentally friendly farming by reducing agrochemical inputs [14]. Farmers can be made accountable for impacting the environment by developing tight regulations and policies. Many researchers have studied the impacts of precision agriculture technologies on the environment [3,15—19] especially nitrogen which is one of the main and essential nutrients for crop growth and production. Over or unnecessary application of nitrogen can cause environmental pollution through surface water, groundwater, and air pollution (Fig. 12.1).

A main concern regarding environmental pollution from agricultural lands is leaching of excessive nutrients to the groundwater [15,17,21,22]. Higher nitrate concentrations in drinking water can potentially cause stomach diseases, cancer, although there are a very less evidence [23,24]. However, excessive amounts of

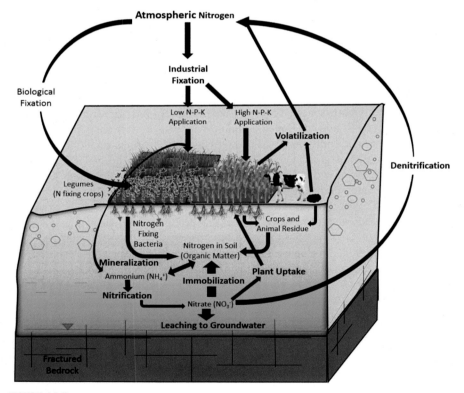

FIGURE 12.1

Different components of the nitrogen cycle and pathways within an agricultural production system [20].

nitarte can lead to methmethemoglobinemia, which causes new-born to turn blue due to a lack of oxygen [23,25]. The PATs in different cropping systems have significantly reduced negative impacts on groundwater quality [17,26]. Eutrophication can be caused by the surface runoff from the agricultural lands transporting large amounts of nutrients, especially phosphorus both in dissolved and particulate forms to freshwater bodies [27]. 4R nutrient stewardship can help to overcome the challenge of excessive nutrients loss in surface runoff [28]. Greenhouse gases (GHGs) emissions are aslo a great environmental concern [29]. The adoption of PATs through less tillage, reduced nutrients application, less fuel consumption, and overall reduction of farm inputs can significantly reduce GHGs emission to the environment. The combination of advancements in PATs development with application of AI algorithms in agriculture decision making will help scientists and farmers to develop and implement decision support systems for early warning applications for weather and disease/pests impacts. This could lead to reducing the negative impacts of the agriculture industry on the environment through enforcing and adopting mitigating strategies to minimize the effects of climate change on crop production.

12.2 Environmental impacts of precision agriculture techniques

Farmers control plant diseases, pests, and weeds primarily through use of agrochemicals, adhering to conventional crop protection strategies, which involve the use of a significant quantity of chemicals [30]. Integrated Pest Management (IPM) is an alternative method of crop protection that alters the philosophy of conventional crop protection by placing a greater emphasis on the understanding of the insect, pest, and crop ecology [31]. IPM also implements several techniques, which in turn reduce risks to human health as well as to the environment [31]. Uniform application of agrochemicals doesn't consider crop and soil variability and excessive application could result in detrimental environmental impacts, higher production costs, and food contamination [32]. However, due to sophisticate sensing mechanisms, precision sprayers can spray on weeds/pests targets and avoid unnecessary application on crop areas. Applications of agrochemicals in targeted areas can significantly reduce production costs, reduce crop damage, minimize chemical traces in food items, and reduce negative impacts on the environment [33].

Numerous studies have been conducted to investigate the effects that various PATs on crop and soil management, farm profitability, crop yields, production costs, groundwater and surface water contamination [2,16,17,26]. Swinton and Lowenberg-DeBoer [34] state that precision agriculture has contributed around 57% to the profitability of the industry. Another study revealed that PATs led to an increase in output as well as a reduction in the costs of inputs, which resulted in increased profitability for farms [35]. Accurate auto guidance systems have the potential to boost crop productivity while increasing farm profitability [36]. Autosteer systems could save farmers 5%—15% of the costs of input (fuel, pesticides, and fertilizer) by reducing over- or under application and efficient time management since farmers can use the autosteer technology both during the day and night [36]. Esau et al. [37] tested and evaluated the smart sprayer in wild blueberry cropping system and results revealed that precise and accurate application of agrochemicals can boost savings of up to 70% and could reduce pollution. Saleem et al. [16] tested and evaluated a variable rate fertilizer spreader and saved 40% of fertilizer as compared to uniform application. This reduced nutrient leaching and phosphorus loss in surface water. Other PATs besides smart spraying mechanisms and autosteer systems also have potential to reduce production costs and reduce environmental pollution. For example PATs have the potential to lessen the risk of agricultural chemicals contaminating ground or surface water [12].

12.2.1 Surface water

Only approximately 3% of the water on the planet is freshwater, whereas the other 97% of the water is saltwater [38]. Despite the fact that two-thirds ($\sim$51M km^3) of the world's freshwater is found in frozen form in the form of polar ice caps and glaciers [39], just one-third of the world's freshwater is stored in rivers, lakes, or underground [40]. In the majority of ecnomically developing nations, 70% of the freshwater used for irrigation goes toward supporting crops [41]. The agriculture industry also contributes to surface water, groundwater, crop uptake, or evaporates into the atmosphere [41]. However, the application of muncipal wastewater can cause soil degradation, waterlogging of irrigated land [42]. Some farmers use wastewater and contaminated water for irrigation that not only have negative impacts on food quality but also contaminate surface and groundwater resources [43]. In the recent years, many worldwide organizations have set the standards for water use in agriculture sector such as the European Nitrate Directive [44], the European Water Directive [45], or the Common Agriculture Policy [46]. Surface runoff from agricultural lands contains both dissolved and particulate nutrients in it which can flow into the rivers or lakes [47]. Elevated amounts of phosphorus (P) in surface water can cause eutrophication [47,48]. Eutrophication can cause oxygen depletion in water which harmfully impacts aquatic life [48]. Making effective use of freshwater when irrigating arable land is crucial factor to consider cutting costs (such as those related to electricity and time) while simultaneously increasing crop yields.

Agriculture is the primary contributor to surface water pollution, in the United States, where frequent monitoring and reporting on the state of water quality in rivers, lakes/reservoirs, and coastal estuaries is conducted [48]. Adoption of PATs and site-specific nutrient management is essential to reduce the surface contamination. Saleem et al. [16] found that variable rate fertilization has reduced the surface water pollution in a wild blueberry cropping system. The field was divided into two sections, uniform and variable rate, and results revealed that excessive nutrients caused increased vegetative growth of wild blueberries in the uniform rate fertilization section with almost double amount of phosphorus in surface water samples collected in the low lying areas of the field [16]. A study in a maize field in New Zealand suggested that higher elevated soil was more susceptible to nutrient runoff and erosion during rainfall events and plant growth was visually less in the higher elevation areas due nutrient runoff [49]. Substantial maize yield improvements were observed using prescription map-based variable rate fertilization (8.4 tons/ha) when compared with traditional uniform application of fertilizers (2.0 tons/ha) in research conducted in central Africa [50]. Kumar et al. [51] suggested that delineation of management zones (MZs) can significantly increase corn yield while reducing nutrients leaching and runoff in irrigated corn fields in Tennessee, USA. Another study conducted in maize fields in Minnesota, USA, revealed that nutrient losses were up to 33% less in variable rate strategy compared to uniform application [8]. In Australia, the sugarcane industry is blamed for nutrient-rich surface runoff that contributes to decline of the Great Barrier Reef, since 95% farms around the reef are sugarcane farms. There has been scrutiny on the industry's practices, namely nutrient-rich runoff or leachate from sugarcane farms [52]. Overall, results of studies conducted in various cropping systems suggested that site-specific nutrient management has huge potential to increase the crop yield and minimize the nutrient loading in surface runoff. However, large-scale studies must be conducted at regional or watershed scale, especially comparing areas with site-specific nutrient management practices, with watersheds under traditional fertilization approaches.

12.2.2 Groundwater

Groundwater (43%) is often used for irrigation, particularly during times of drought when the crop water requirements are at their highest [53,54]. However, continuous pumping of groundwater can increase the risks of groundwater depletion, which can not only put the future groundwater supply at danger [55], but can also lower river base flow and have negative effects on aquatic life [55]. Therefore, it is of utmost importance to preserve both the quality and the quantity of groundwater that is stored in aquifers in order to guarantee a reliable supply of potable water for future generations. One of the most widespread problems in every region of the world is pollution of the groundwater supply through anthropogenic activities [56]. Urbanization, industrial growth, intensive agriculture, and food processing in the last four to 5 decades has led to an increase in the demand of available water resources [57]. The use of nitrogen fertilizers, which are very inexpensive and allow for a large increase in crop yield, has been a driving force behind the intensification of agriculture [58]. When nitrogen fertilizers are applied in quantities that are greater than what the plants require, nitrogen and salt surpluses typically drain into groundwater bodies [59]. Nitrogen fertilizers can be either organic or mineral in nature; however, mineral nitrogen is more generally used [60]. The most significant contributor to the presence of nitrogen in groundwater is agriculture, specifically the excessive application of fertilizer [61]. Higher nitrate concentrations in drinking water can potentially cause cancer, although there is very less evidence [62]. As noted previously several nations and supranational organizations have come up with plans and measures of protection and mitigation to bring under control the excessive nitrogen concentrations that have been found in surface and groundwater [25].

Transformation of forested lands into agricultural lands in China have increased the nitrate concentrations in groundwater [63]. In 2004, the concentrations of nitrate in the groundwater in the largely agricultural Xiaojiang river basin of Yunnan province were found to be high above the criteria for acceptable levels of nitrate in drinking water [63]. The study results suggested that the amount of land used for agriculture has increased by 132.7% over the past 2 decades. The shift in the chemical composition of groundwater is related to the conversion of previously forested land to agricultural land [63]. Saleem [20] presented results of various studies conducted in Canada on groundwater contamination from agricultural lands and their findings suggested that a significant proportion (11 to 55%) of groundwater wells had nitrate concentrations above the allowable drinking water limits (Table 12.1).

Water table depth, geographic formations, topsoil, crop type, and climate variability are considered the most influential factors impacting the nitrate concentrations in groundwater [64–67]. A combination of thin overburden and shallow groundwater presence can significantly increase the chances of nitrate contamination of groundwater especially in agricultural areas where fertilizers are applied with irrigation water [68,69]. Groundwater nitrate concentrations can also be influenced by crop type in agricultural watersheds [70] since fertilizer requirements for different cropping systems vary depending upon the nitrogen demand of the crop and the time of fertilization [71]. Rotations of high nitrogen demanding crops, such as corn and wheat, with different low nitrogen demanding crops such as soybeans, peas, lentil which have lowered nitrate leaching rates [72]. For example, a corn-soybean-rye rotation has the potential to decrease nitrate leaching from fields compared to successive corn plantings [20,73]. For example, reduction in nitrogen application rates from 168 to 134 kg/ha and from 134 to 112 kg N/ha using a corn-soybean rotation in Iowa lowered the nitrate concentrations in drainage water by 10% and 20%, respectively [74]. A 15-year model simulation indicated that a variable rate nitrogen split side-dress treatment reduced N load by an average of 40% compared to a single preplant application [74]. Modifying the timing and rate of N application decreased the quantity of residual soil N, but had no effect on the amount of soil organic matter mineralized, as predicted by the model [75]. However, crop rotation can also result in higher nitrate leaching and the risk of nitrate leaching from these crops is also dependent on soil and environmental conditions such as fertility status of the soil and dry or wet weather conditions [76].

Variable rate technologies (VRTs) studies were conducted in Czech Republic, and their results suggested that VRTs have huge potential to lower the groundwater contamination as well as increase farm profitability by 10 to15% [26]. A study conducted in central China compared the effects of precision seeding and precision land leveling with traditional farming. The results suggested that maize production was increased by 10% while there was significant decrease in nitrate

Table 12.1 Previous Canadian studies about nitrate-N (mg/L) presence in groundwater (without spatial coordinates) [20].

Study	Year	Province	No. of sites	% of sites above 10 mg/L
Johnston (1955)	1950–54	Ontario	484	13.8
Hill (1982)	1980–81	Ontario	164	40
Novakovic (1985)	1984–85	Ontario	63	21
Howard and Falck (1986)	1985	Ontario	45	25
Ecobichon et al. (1990)	1985–86	New Brunswick	300	20.2
Richards et al. (1990)	1988	New Brunswick	47	20
Frank et al. (1991)	1986–87	Ontario	183	21
Lee-Han and Hatton (1991)	1991–92	Ontario	566	11.7
Fleming (1992)	1991	Ontario	301	15
Wei et al. (1993)	1989	British Columbia	100	23
Moerman and Briggins (1994)	1989–93	Nova Scotia	237	12.6
Goss and Goorahoo (1995)	1991–92	Ontario	1300	14
Wassenaar (1995)	1993	British Columbia	117	54
Goss et al. (1998)	1991–92	Ontario	1212	14
Rudolph et al. (1998)	1991–92	Ontario	137	23
Thompson (2001)	1995–2000	Saskatchewan	3425	13.5
Bonton et al. (2010)	2005–07	Quebec	837	40

concentrations in subsurface water [77]. Saleem et al. [17] a study in the wild blueberry suggested that there was insignificant difference in berry yield yet, groundwater nitrate concentrations were significantly lowered in variable rate fertilization compared to uniform fertilization. Bohman et al. [78] suggested that reducing the irrigation rate by 15% could reduce the nitrate loading in groundwater by 17% in potato field. Variable rate fertilization has the potential to increase crop yield when compared with traditional practices. Implementation of groundwater conservation techniques such as best management practices, crop rotations, and variable rate nutrient management are essential to lower the nitrate concentrations and make agriculture sustainable considering the future climate change and human dependency on groundwater resources for both drinking and irrigation purposes.

12.2.3 Air

The consumption of conventional fertilizers like urea, nitrogen, phosphorus, and potassium has been gradually increasing over the past several years with the intention of increasing crop yield in the agriculture and agrofood sectors [79]. Usually, chemical fertilizer has a low fertilizer efficiency because of its high solubility, low thermal stability, and small molecular weight [80]. This causes the majority of the nutrients in the fertilizer to be lost to the environment through surface runoff, denitrification, leaching, and volatilization [80]. Organic fertilizer such as farm yard manure has a higher efficiency because it has a higher molecular weight and a lower solubility [81]. In addition, the usage of conventional fertilizer is expensive because of the low nutrient utilization efficiency (NUE) that it possesses [81]. Therefore, controlled fertilizer such as biochar that has mechanism for prolonged release is benefical for environmental protection. Over the course of the past 10 years, numerous controlled release fertilizers have been the subject of extensive research since they have the potential to improve fertilizer efficiency, as well as nutrient retention, by lowering the solubility of the nutrient in question, and dispersing the nutrient in a manner that is under the operator's direct control, while at the same time ensuring that the plant receives the essential nutrients for healthy development [82]. In recent years, the use of biochar as an ingredient in the formulation of fertilizer has become a popular new trend [83]. In addition to organic amendments, in agricultural field operations, the implementation of precision agriculture (PA) practices that make use of a large reservoir of PATs can make a positive contribution to the reduction of greenhouse gas emissions for the following reasons: (i) the enhancement of the capability of soils to function as carbon stock reserve [84] by less tillage [84] and reduced nitrogen fertilization [85,86]; (ii) the reduction of fuel consumption due to fewer in-field operations with the tractor. Therefore, Greenhouse Gases (GHGs) mitigation measures, which refer to new technologies and techniques on all agricultural practices (precision/variable rate sowing/planting, fertilizing, spraying, and irrigation) can significantly reduce the amount of inputs that are responsible for the contribution of GHGs and could help reduce the climate change impact of agriculture. However, it is important to keep in mind that crop production should be maintained or even increased in order to meet the challenge of ensuring food security.

Balafoutis et al. [33] suggested that precision agriculture technologies have positively reduced the GHGs emissions from agricultural fields in Europe. A study in rice field suggested that PATs not only have potential to reduce the GHGs emissions but also reduce the cost of production [80]. Abbas et al. [87] tested the variable rate split fertilization in wild blueberry fields and found that split variable rate fertilization can significantly reduce the volatilization losses. Overall, most of the studies have showed that PATs adoption in various cropping systems have positive impacts on surface water, groundwater, and air.

12.3 Potential climate change impacts

Climate change is inducing abrupt changes in the hydrological cycle, including rainfall and drought, and in extreme high and low temperature which have a significant influence on society as well as the natural environment. Globally, an increase in the intensity and frequency of flood events is a major concern over the last three decades as it is the most wide spread natural hazard [88,89]. Flooding can cause major decline in the yield of the cereal crops, it could be more severe in many parts of the world due to climatic anomaly in the future [90]. Climate change also effect the ecosystem and changes in temperature and rainfall may caused distribution of diease vactors, e.g., those of malaria and dengue, and the incidence of diarrhoeal diseases. It also caused the air pollutants, for example tropospheric ozone pollution which was high in European areas. Sea level rise is likely to threaten low lying coastal populations, particularly in countries where economic conditions don't allow sea defense may lead population displacement and more environmental refugees [91,92]. Therefore, there is special guidance accessible to flood management authorities as well as local planners, with the goal of assisting them in managing flood risk in a manner that considers the potential implications of climate change [93]. The scientific literature presents several different potential outcomes regarding climate change. However, the Intergovernmental Panel on Climate Change (IPCC) refers to four distinct scenarios in its fifth Assessment Report (AR5). These scenarios are referred to as RCP 2.6 (Representative Concentration Pathway 2.6), RCP 4.5, RCP 6.0, and RCP 8.5 (Intergovernmental Panel on Climate Change [IPCC], 2013; and IPCC, 2014). When compared to the year 1750, which stands for the preindustrial levels, the quantity that identifies each scenario approximates the Radiative Forcing (RF), expressed in W/m^2, either at the year 2100 or at stabilization afterward. The amount of change in the energy flux per unit of surface area is referred to as the radiative forcing. Positive radiative forcing is responsible for warming the surface, and negative radiative forcing is responsible for cooling the surface [94].

Fader et al. [95] evaluated the potential impact that climate change and rising CO_2 concentrations in the atmosphere may have on the need for irrigation in the Mediterranean region, taking into account the effects of shifting demographics and advances in agricultural technology. When applied to the Mediterranean region, certain climate models at global warming of 3°C and above showed a signal of increasing the net irrigation requirements. This was the case even though higher CO_2 concentrations in the atmosphere had the potential to have beneficial effects. Rolim et al. [96] expected that there will be an increase in the need for water in irrigation systems so that agricultural yields can remain at their current levels. According to Kakumanu et al. [97], in recent years, water resources, agriculture, ecology, and other fields have become hotspots for research due to the conditions of climate change caused by global warming. These conditions are characterized by the fact that the Earth's temperature has increased. The availability of water resources and the agricultural food production system are both being negatively impacted by climate change in many developing countries [98]. According to the findings of studies, there appears to be a tendency toward less rainfall and a trend toward higher surface temperatures. Through the Climate Adapt program, several different adaption techniques were created and put into action in order to alleviate the effects of climate change. Bocci and Smanis [99] indicated that for all southern Mediterranean countries, changes in temperature and precipitation patterns that were predicted by a general atmospheric circulation model are already affecting the sector by increasing the sector's exposure to risks of floods and extreme droughts.

12.3.1 Extreme events and role of precision agriculture

The increase in the frequency of extreme weather events which are related to climate change represents a severe threat to agriculture and for the farmer [100]. Growers have already experienced the seasonal shift and suffer with seasonal food insecurity [101]. Climatology are depend on the projections made by global climate models, also known as GCMs, because of the lengthy lag time before observations can indicate a new climate [102]. The statistical measure of the "change" in climatic conditions, such as the range of warm to cool annual maximum temperatures over the course of years, is referred to as climatic variability [103]. A climatic anomaly is the departure of a specific climatic state from both its average and its variability [104]. Extreme events that has the most significant influence on the growth and development of the crop especially fluctuations in regional crop production [105]. There is some evidence that precision agriculture, or more specifically precision technologies, may play a role in mitigate the climate change as the focus of PATs is to reduce the production cost and environmental effects to increase the farm's profitability [106]. In PATs, IoT based smart sensors are used in the agriculture field for collecting data related to soil nutrients, fertilizers, and water requirements as well as for analyzing the crop growth [107]. Firstly, a grower need to maintain accurate records of weather occurrences as well as the impact these occurrences have on crop yields in the current climate. The second step for the same farmer is to examine the forecasts of climate change for their region, and then search for observations made in a different region of the world that is presently enduring conditions that are analogous to the new climate. The producer will be able to foresee the future influences on output in the new environment if they have a thorough awareness of the ways in which crop growth and yields are affected by the existing climate in the other geography.

Growers can progressively make changes in their management decisions, such as the choice of seed, the time of planting, and other management decisions, if they have knowledge of how the future trend in climate will affect the yield of their crops. Since climate trends can exacerbate local weather patterns, affecting crop productivity, this precision monitoring needs to extend across the entirety of a farm's area. For instance, in a future climate condition with fewer occurrences of precipitation, dry soils that currently exist in one section of a farm may also occur in previously unaffected areas. Moreover, early season weather that is warmer in the future may result in conditions that are conducive for the development of disease throughout life stages of a crop that are more sensitive to it. A farmer may plant a crop earlier in the season due to warmer temperatures, but they face the risk of a late frost. The recent patterns of drier weather across a large portion of the Midwest USA, along with temperature highs that broke records over the course of this summer, should serve as fair warning that climate variability is real regardless of the source [108]. Precision agriculture, in particular innovative approaches such as variable-rate irrigation, can be a part of the solution to help reduce the risk that is linked with the potential climate conditions of the future. It is simply a matter of comprehension and putting the plan into action.

12.3.2 Nutrient management based on spatiotemporal climate

The adoption of PATs is a relatively recent development and helps farmers to develop an understanding of the crop's requirements for a profitable yield, the efficiency with which the crop is utilized, the capacity of the soil to supply a portion of the crop's requirements, and the temporal patterns of crop uptake and utilization in comparison to supply from the soil volume are required for the application of nutrients [109,110]. The use of nutrients presents a chance for the efficient management of inputs, which, in turn, helps to boost the effectiveness of output. The primary objective of PA is to maximize output with a minimum of resources used while simultaneously lowering environmental impact [111,112]. Precision agriculture with the goal of inputs management will provide differentiated production methods for agricultural producers, and similar to any other technology, it may enable farmers to collect data to uncover effective variables on their farm's potential yield. In addition, farmers can make choices regarding inputs and employ them in varying proportions [113].

12.4 Challenges for farmers and researchers

The choice of whether to engage in environmentally responsible farming because of financial incentives offered to farmers is not an either/or proposition. Adoption is contingent on a wide variety of parameters, including the requirements of the programmer, the incentives that are provided, the preferences of the environment, the individual perspectives, experiences, and education of farmers [112,114]. The programs [115,116] of the incentive kind that are related to short-term economic advantage have a higher adoption rate than programs that are exclusively focused on providing an environmental service. One of the most powerful drivers for farmers to embrace environmentally friendly farming methods is the belief that doing so will benefit either their farms the environment or both in the long run. Beside financial benefits, there is a heavy emphasis placed on the importance of extension services and technical help in the process of spreading sustainable practices. After looking at a variety of studies [117,118], researchers found that there are three distinct categories of incentives: market and nonmarket incentives, regulatory measures, and cross-compliance measures that link direct payments to farmers' compliance with fundamental environmental standards or the maintenance of land in good agricultural and environmental condition. The author [116] has been demonstrated in a number of literature reviews that financial rewards that encourage better business practices are more likely to be adopted in the near term, particularly if they are undertaken on a voluntarily basis. However, over the course of many years, primary motivators will be favorable outcomes for either the farm or the ecosystem.

12.5 Conclusions

Future challenges in crop production will remain similar to what growers and producers are facing today. Environmental challenges such as climate change, ground and surface water pollution, and water shortage will not be considered policy issues until the scientific community is able to sufficiently identify them. The developed countries have adopted modern technologies to enhanced crop growth and yields and increased the farm inputs. While developing countries are still using traditional crop production mechanisms. However, the traditional farming practices will become obsolete due to continue threat of climate change. For example, crop production using traditional techniques will be replaced with smart farming techniques to feed global population. A comprehensive understanding of environment will help us to mitigate the harmful and determent effects excessive use of agrochemical on the environment. Development of PATs in the last couple of decades has reduced the cost of production and reduced environmental pollution. However, more focus is required to adopt these PATs, AI applications, crop rotation, and climate adoptive crops in the next few decades. Threats of climate change may force farmers to adopt these technologies at faster rates which could be beneficial to the environment. Scientists and researchers should also focus on developing farmer-friendly and customized technologies to boost crop production and save the environment.

References

[1] Conway GR, Barbie EB. After the green revolution: sustainable and equitable agricultural development. Futures 1988;20(6):651–70.
[2] Scudamore KA, Hazel CM, Patel S, Scriven F. Deoxynivalenol and other *Fusarium mycotoxins* in bread, cake, and biscuits produced from UK-grown wheat under commercial and pilot scale conditions. Food Addit Contam 2009;26(8):1191–8.
[3] Wilkins RJ. Eco-efficient approaches to land management: a case for increased integration of crop and animal production systems. Phil Trans Biol Sci 2008;363(1491): 517–25.

[4] Zheng W, Luo B, Hu X. The determinants of farmers' fertilizers and pesticides use behavior in China: an explanation based on label effect. J Clean Prod 2020;272: 123054.

[5] Lee WS, Alchanatis V, Yang C, Hirafuji M, Moshou D, Li C. Sensing technologies for precision specialty crop production. Comput Electron Agric 2010;74(1):2–33.

[6] Kumar R, Kumar R, Prakash O. Chapter-5 the impact of chemical fertilizers on our environment and ecosystem. Chief Ed 2019;35:69.

[7] Zhang F, Cui Z, Fan M, Zhang W, Chen X, Jiang R. Integrated soil–crop system management: reducing environmental risk while increasing crop productivity and improving nutrient use efficiency in China. J Environ Qual 2011;40(4):1051–7.

[8] Mulla DJ, Schepers JS. Key processes and properties for site-specific soil and crop management. The state of site specific management for agriculture, vols. 1–18; 1997.

[9] Whitmeyer SJ, Nicoletti J, de Paor DG. The digital revolution in geologic mapping. GSA Today (Geol Soc Am) 2010;20(4/5):4–10.

[10] O'Hanlon ME. Neither Star Wars nor sanctuary: constraining the military uses of space. Brookings Institution Press; 2004.

[11] Bongiovanni R, Lowenberg-DeBoer J. Precision agriculture and sustainability. Precis Agric 2004;5(4):359–87.

[12] Far ST, Rezaei-Moghaddam K. Impacts of the precision agricultural technologies in Iran: an analysis experts' perception & their determinants. Inf Process Agric 2018; 5(1):173–84.

[13] Kritikos M. Precision agriculture in Europe: legal and ethical reflections for lawmakers. European Parliament's Science and Technology Options Assessment (STOA); 2017.

[14] Ahirwar NK, Singh R, Chaurasia S, Chandra R, Ramana S. Effective role of beneficial microbes in achieving the sustainable agriculture and eco-friendly environment development goals: a review. Front Microbiol 2020;5:111–23.

[15] Saleem S, Levison J, Parker B, Martin R, Persaud E. Impacts of climate change and different crop rotation scenarios on groundwater nitrate concentrations in a sandy aquifer. Sustainability 2020;12(3):1153.

[16] Saleem SR, Zaman QU, Schumann AW, Madani A, Chang YK, Farooque AA. Impact of variable rate fertilization on nutrients losses in surface runofffor wild blueberry fields. Appl Eng Agric 2014;30(2):179–85.

[17] Saleem SR, Zaman QU, Schumann AW, Madani A, Farooque AA, Percival DC. Impact of variable rate fertilization on subsurface water contamination in wild blueberry cropping system. Appl Eng Agric 2013;29(2):225–32.

[18] Kumar S, Karaliya SK, Chaudhary S. Precision farming technologies towards enhancing productivity and sustainability of rice-wheat cropping system. Int J Curr Microbiol App Sci 2017;6(3):142–51.

[19] Kassam A, Brammer H. Environmental implications of three modern agricultural practices: conservation agriculture, the system of rice intensification and precision agriculture. Int J Environ Stud 2016;73(5):702–18.

[20] Saleem SR. Impacts of future climate and agricultural land use changes on groundwater nitrate concentrations in southern Ontario. 2018.

[21] Craswell E. Fertilizers and nitrate pollution of surface and ground water: an increasingly pervasive global problem. SN Appl Sci 2021;3(4):1–24.

[22] Li P, Karunanidhi D, Subramani T, Srinivasamoorthy K. Sources and consequences of groundwater contamination. Arch Environ Contam Toxicol 2021;80(1):1–10.

[23] Johnson SF. Methemoglobinemia: infants at risk. Curr Probl Pediatr Adolesc Health Care 2019;49(3):57—67.

[24] Weyer PJ, et al. Municipal drinking water nitrate level and cancer risk in older women: the Iowa Women's Health Study. Epidemiology 2001:327—38.

[25] Smith RP. What makes my baby blue. Dartmouth Medicine 2000:26—31.

[26] Fabiani S, Vanino S, Napoli R, Zajíček A, Duffková R, Evangelou E, et al. Assessment of the economic and environmental sustainability of variable rate technology (VRT) application in different wheat intensive European agricultural areas. A water energy food nexus approach. Environ Sci Pol 2020;114:366—76.

[27] Martín M, Hernández-Crespo C, Andrés-Doménech I, Benedito-Durá V. Fifty years of eutrophication in the Albufera lake (Valencia, Spain): causes, evolution and remediation strategies. Ecol Eng 2020;155:105932.

[28] Sela S, van Es HM. Dynamic tools unify fragmented 4Rs into an integrative nitrogen management approach. J Soil Water Conserv 2018;73(4):107A.

[29] Hou D, Ding Z, Li G, Wu L, Hu P, Guo G, et al. A sustainability assessment framework for agricultural land remediation in China. Land Degrad Dev 2018;29(4):1005—18.

[30] Hillocks RJ. Farming with fewer pesticides: EU pesticide review and resulting challenges for UK agriculture. Crop Protect 2012;31(1):85—93.

[31] Lamichhane JR, Aubertot JN, Begg G, Birch ANE, Boonekamp P, Dachbrodt-Saaydeh S, et al. Networking of integrated pest management: a powerful approach to address common challenges in agriculture. Crop Protect 2016;89:139—51.

[32] Lu Y, Song S, Wang R, Liu Z, Meng J, Sweetman AJ, et al. Impacts of soil and water pollution on food safety and health risks in China. Environ Int 2015;77:5—15.

[33] Balafoutis A, Beck B, Fountas S, Vangeyte J, van der Wal T, Soto I, et al. Precision agriculture technologies positively contributing to GHG emissions mitigation, farm productivity and economics. Sustainability 2017;9(8):1339.

[34] Lowenberg-DeBoer J, Swinton SM. Economics of site-specific management in agronomic crops. In: The state of site specific management for agriculture. (ASA, CSSA, and SSSA Books); 1997. p. 369—96. https://doi.org/10.2134/1997.stateofsitespecific.c16.

[35] Bullock DS, Lowenberg-DeBoer J, Swinton SM. Adding value to spatially managed inputs by understanding site-specific yield response. Agric Econ 2002;27(3):233—45. https://doi.org/10.1111/j.1574-0862.2002.tb00119.x.

[36] Schieffer J, Dillon C. The economic and environmental impacts of precision agriculture and interactions with agro-environmental policy. Precis Agric 2015;16(1):46—61. https://doi.org/10.1007/s11119-014-9382-5.

[37] Esau T, Zaman Q, Groulx D, Farooque A, Schumann A, Chang Y. Machine vision smart sprayer for spot-application of agrochemical in wild blueberry fields. Precis Agric 2018;19(4):770—88. https://doi.org/10.1007/s11119-017-9557-y.

[38] Kalogirou S. Survey of solar desalination systems and system selection. Energy 1997;22(1):69—81. https://www.sciencedirect.com/science/article/pii/S0360544296001004.

[39] Irvine-Fynn TD, Edwards A. A frozen asset: the potential of flow cytometry in constraining the glacial biome. Cytometry part A 2014;85(1):3—7.

[40] Bárdossy A, el Hachem A. Assessment of water quantity. In: Bogardi JJ, Gupta J, Nandalal KDW, Salamé L, van Nooijen RRP, Kumar N, et al., editors. Handbook of water resources management: discourses, concepts and examples. Cham: Springer International Publishing; 2021. p. 443—69. https://doi.org/10.1007/978-3-030-60147-8_14.

[41] Wang X. Managing land carrying capacity: key to achieving sustainable production systems for food security. Land 2022;11(4):484.

[42] Hallberg GR. Agricultural chemicals in ground water: extent and implications. Am J Alternative Agric 1987;2(1):3–15. https://www.cambridge.org/core/article/agricultural-chemicals-in-ground-water-extent-and-implications/04E60320AE6B7B8474C3D9ADC14EEADD.

[43] Qadir M, Wichelns D, Raschid-Sally L, McCornick PG, Drechsel P, Bahri A, et al. The challenges of wastewater irrigation in developing countries. Agric Water Manag 2010;97(4):561–8. https://www.sciencedirect.com/science/article/pii/S0378377408002989.

[44] Brouwer F, Hellegers P. The nitrate directive and farming practice in the European Union. European Environ 1996;6(6):204–9. https://doi.org/10.1002/(SICI)1099-0976(199611)6:6<204::AID-EET93>3.0.CO.

[45] Mostert E. The European water framework directive and water management research. Phys Chem Earth Parts A/B/C 2003;28(12):523–7. https://www.sciencedirect.com/science/article/pii/S1474706503000895.

[46] Ackrill R, Studies UACE. Common agricultural policy. Bloomsbury Academic; 2000 (Contemporary European Studies). Available from: https://books.google.com.pk/books?id=qbjUAwAAQBAJ.

[47] Schoumans OF, Chardon WJ, Bechmann ME, Gascuel-Odoux C, Hofman G, Kronvang B, et al. Mitigation options to reduce phosphorus losses from the agricultural sector and improve surface water quality: a review. Sci Total Environ 2014;468–469:1255–66. https://www.sciencedirect.com/science/article/pii/S0048969713009881.

[48] Dorgham MM. Effects of eutrophication. In: Ansari AA, Gill SS, editors. Eutrophication: causes, consequences and control, vol 2. Dordrecht: Springer Netherlands; 2014. p. 29–44. https://doi.org/10.1007/978-94-007-7814-6_3.

[49] Jiang G, Grafton M, Pearson D, Bretherton M, Holmes A. Integration of precision farming data and spatial statistical modelling to interpret field-scale maize productivity. Agriculture 2019;9(11):237 [cited 2022 Nov 9]; Available from: www.mdpi.com/journal/agriculture.

[50] Ulimwengu JM, Kibonge A. Soil mapping, fertilizer application, and maize yield: a spatial econometric approach working. 2022. Paper. [cited 2022 Nov 9]; Available from: www.akademiya2063.org.

[51] Kumar H, Srivastava P, Lamba J, Ortiz Bv, Way TR, Sangha L, et al. Phosphorus variability in the irrigated cropland during a growing season. In: 2021 ASABE annual international virtual meeting, July 12–16, 2021. American Society of Agricultural and Biological Engineers; 2021.

[52] Bell M, Schaffelke B, Moody P, Waters D, Silburn M. Tracking nitrogen from the paddock to the reef—a case study from the great barrier reef. 2016 [cited 2022 Nov 9]; Available from: www.ini2016.com.

[53] Drechsel P., Heffer P., Magen H., Mikkelsen R., Singh H., Wichelns D. Managing water and nutrients to ensure global food security, while sustaining ecosystem services. 2015.

[54] Dubois O. The State of the World's Land and Water Resources for Food and Agriculture: Managing Systems at Risk. Earthscan; 2011.

[55] Gleeson T, Richter B. How much groundwater can we pump and protect environmental flows through time? Presumptive standards for conjunctive management of aquifers and rivers. River Res Appl 2018;34(1):83–92.

[56] Li P, Tian R, Xue C, Wu J. Progress, opportunities, and key fields for groundwater quality research under the impacts of human activities in China with a special focus on western China. Environ Sci Pollut Control Ser 2017;24(15):13224—34.

[57] Cai X, Rosegrant MW. Optional water development strategies for the Yellow River Basin: balancing agricultural and ecological water demands. Water Resour Res 2004; 40(8).

[58] Baethgen WE. Vulnerability of the agricultural sector of Latin America to climate change. Clim Res 1997;9(1—2):1—7.

[59] Khan MN, Mohammad F. Eutrophication: challenges and solutions. In: Eutrophication: causes, consequences and control. Springer; 2014. p. 1—15.

[60] Dion PP, Jeanne T, Thériault M, Hogue R, Pepin S, Dorais M. Nitrogen release from five organic fertilizers commonly used in greenhouse organic horticulture with contrasting effects on bacterial communities. Can J Soil Sci 2020;100(2):120—35.

[61] Johnson G v, Raun WR. Nitrate leaching in continuous winter wheat: use of a soil-plant buffering concept to account for fertilizer nitrogen. J Prod Agric 1995;8(4): 486—91.

[62] Weyer PJ, Cerhan JR, Kross BC, Hallberg GR, Kantamneni J, Breuer G, et al. Municipal drinking water nitrate level and cancer risk in older women: the Iowa Women's Health Study. Epidemiology 2001:327—38.

[63] Jiang Y, Zhang C, Yuan D, Zhang G, He R. Impact of land use change on groundwater quality in a typical karst watershed of southwest China: a case study of the Xiaojiang watershed, Yunnan Province. Hydrogeol J 2008;16(4):727—35. https://doi.org/10.1007/s10040-007-0259-9.

[64] Levison J, Novakowski K. The impact of cattle pasturing on groundwater quality in bedrock aquifers having minimal overburden. Hydrogeol J 2009;17(3):559—69.

[65] Seidenfaden IK, Sonnenborg TO, Børgesen CD, Trolle D, Olesen JE, Refsgaard JC. Impacts of land use, climate change and hydrological model structure on nitrate fluxes: magnitudes and uncertainties. Sci Total Environ 2022;830:154671. https://www.sciencedirect.com/science/article/pii/S0048969722017648.

[66] Lin X. Promoting the sustainable utilization of groundwater resources in Ethiopia using the integrated groundwater footprint index. 2020.

[67] Dixit A, Madhav S, Mishra R, Srivastav AL, Garg P. Impact of climate change on water resources, challenges and mitigation strategies to achieve sustainable development goals. Arabian J Geosci 2022;15(14):1296. https://doi.org/10.1007/s12517-022-10590-9.

[68] Baram S, Couvreur V, Harter T, Read M, Brown PH, Hopmans JW, et al. Assessment of orchard N losses to groundwater with a vadose zone monitoring network. Agric Water Manag 2016;172:83—95.

[69] McAleer E, Coxon C, Mellander PE, Grant J, Richards K. Patterns and drivers of groundwater and stream nitrate concentrations in intensively managed agricultural catchments. Water (Basel) 2022;14(9):1388.

[70] Zhang D, Wang P, Cui R, Yang H, Li G, Chen A, et al. Electrical conductivity and dissolved oxygen as predictors of nitrate concentrations in shallow groundwater in Erhai Lake region. Sci Total Environ 2022;802:149879. https://www.sciencedirect.com/science/article/pii/S0048969721049548.

[71] Rathke GW, Behrens T, Diepenbrock W. Integrated nitrogen management strategies to improve seed yield, oil content and nitrogen efficiency of winter oilseed rape (*Brassica napus* L.): a review. Agric Ecosyst Environ 2006;117(2—3):80—108.

[72] Allende-Montalbán R, Martín-Lammerding D, del Mar Delgado M, Porcel MA, Gabriel JL. Nitrate leaching in maize (*Zea mays* L.) and wheat (*Triticum aestivum* L.) irrigated cropping systems under nitrification inhibitor and/or intercropping effects. Agriculture 2022;12(4):478.

[73] Fry JE, Guber AK. Temporal stability of field-scale patterns in soil water content across topographically diverse agricultural landscapes. J Hydrol (Amst). 2020;580: 124260.

[74] Lawlor PA, Helmers MJ, Baker JL, Melvin SW, Lemke DW. Nitrogen application rate effect on nitrate-nitrogen concentration and loss in subsurface drainage for a corn-soybean rotation. Trans ASABE (Am Soc Agric Biol Eng) 2008;51(1): 83−94.

[75] Wilson GL, Mulla DJ, Galzki J, Laacouri A, Vetsch J, Sands G. Effects of fertilizer timing and variable rate N on nitrate−N losses from a tile drained corn-soybean rotation simulated using DRAINMOD-NII. Precis Agric 2020;21(2):311−23. https://doi.org/10.1007/s11119-019-09668-4.

[76] Tei F, de Neve S, de Haan J, Kristensen HL. Nitrogen management of vegetable crops. Agric Water Manag 2020;240:106316.

[77] Wang Y, Liu Y. Benefits of precision agriculture application for winter wheat in Central China. In: 2018 7th international conference on agro-geoinformatics (Agro-geoinformatics); 2018. p. 1−4.

[78] Bohman BJ, Rosen CJ, Mulla DJ. Impact of variable rate nitrogen and reduced irrigation management on nitrate leaching for potato. J Environ Qual 2020;49(2):281−91. https://doi.org/10.1002/jeq2.20028.

[79] Billen G, Aguilera E, Einarsson R, Garnier J, Gingrich S, Grizzetti B, et al. Reshaping the European agro-food system and closing its nitrogen cycle: the potential of combining dietary change, agroecology, and circularity. One Earth 2021;4(6):839−50.

[80] Pathak H. Mitigating greenhouse gas and nitrogen loss with improved fertilizer management in rice: quantification and economic assessment. Nutrient Cycl Agroecosyst 2010;87(3):443−54.

[81] Vishwakarma K, Upadhyay N, Kumar N, Tripathi DK, Chauhan DK, Sharma S, et al. Potential applications and avenues of nanotechnology in sustainable agriculture. In: Nanomaterials in plants, algae, and microorganisms. Elsevier; 2018. p. 473−500.

[82] Dimkpa CO, Fugice J, Singh U, Lewis TD. Development of fertilizers for enhanced nitrogen use efficiency—trends and perspectives. Sci Total Environ 2020;731:139113.

[83] Sim DH, Tan IA, Lim LL, Hameed BH. Encapsulated biochar-based sustained release fertilizer for precision agriculture: a review. J Clean Prod 2021;303:127018.

[84] Angers DA, Eriksen-Hamel NS. Full-inversion tillage and organic carbon distribution in soil profiles: a meta-analysis. Soil Sci Soc Am J 2008;72(5):1370−4. https://doi.org/10.2136/sssaj2007.0342.

[85] Waldrop MP, Zak DR, Sinsabaugh RL, Gallo M, Lauber C. Nitrogen deposition modifies soil carbon storage through changes in microbial enzymatic activity. Ecol Appl 2004;14(4):1172−7. https://doi.org/10.1890/03-5120.

[86] Khan SA, Mulvaney RL, Ellsworth TR, Boast CW. The myth of nitrogen fertilization for soil carbon sequestration. J Environ Qual 2007;36(6):1821−32. https://doi.org/10.2134/jeq2007.0099.

[87] Abbas A, Zaman QU, Schumann AW, Brewster G, Donald R, Chattha HS. Effect of split variable rate fertilizationon ammonia volatilization in wild blueberry cropping system. Appl Eng Agric 2014;30(4):619−27.

[88] Liao KH, Le TA, Van Nguyen K. Urban design principles for flood resilience: learning from the ecological wisdom of living with floods in the Vietnamese Mekong Delta. Landsc Urban Plan 2016;155:69–78.

[89] Wingfield T, Macdonald N, Peters K, Spees J, Potter K. Natural flood management: beyond the evidence debate. Area 2019;51(4):743–51.

[90] Jia W, Ma M, Chen J, Wu S. Plant morphological, physiological and anatomical adaption to flooding stress and the underlying molecular mechanisms. Int J Mol Sci 2021;22(3):1088.

[91] Langner J, Bergström R, Foltescu V. Impact of climate change on surface ozone and deposition of sulphur and nitrogen in Europe. Atmos Environ 2005;39(6):1129–41.

[92] Haines A, Kovats RS, Campbell-Lendrum D, Corvalán C. Climate change and human health: impacts, vulnerability and public health. Public health 2006;120(7):585–96.

[93] Reynard NS, Kay AL, Anderson M, Donovan B, Duckworth C. The evolution of climate change guidance for fluvial flood risk management in England. Prog Phys Geogr 2017;41(2):222–37.

[94] Allen SK, Plattner GK, Nauels A, Xia Y, Stocker TF. Climate change 2013: the physical science basis. An overview of the working group 1 contribution to the fifth assessment report of the intergovernmental panel on climate change (IPCC). In: EGU general assembly conference abstracts; 2014. p. 3544.

[95] Fader M, Shi S, von Bloh W, Bondeau A, Cramer W. Mediterranean irrigation under climate change: more efficient irrigation needed to compensate for increases in irrigation water requirements. Hydrol Earth Syst Sci 2016;20(2):953–73.

[96] Rolim J, Teixeira JL, Catalao J, Shahidian S. The impacts of climate change on irrigated agriculture in Southern Portugal. Irrigat Drain 2017;66(1):3–18.

[97] Kakumanu KR, Kaluvai YR, Nagothu US, Lati NR, Kotapati GR, Karanam S. Building farm-level capacities in irrigation water management to adapt to climate change. Irrigat Drain 2018;67(1):43–54.

[98] Kotir JH. Climate change and variability in Sub-Saharan Africa: a review of current and future trends and impacts on agriculture and food security. Environ Dev Sustain 2011 Jun;13(3):587–605.

[99] Bocci M, Smanis T. Assessment of the impacts of climate change on the agriculture sector in the southern Mediterranean. In: Union for the Mediterranean: Forseen Development and policy measures; 2019. p. 1–36.

[100] Cogato A, Meggio F, De Antoni Migliorati M, Marinello F. Extreme weather events in agriculture: a systematic review. Sustainability 2019;11(9):2547.

[101] Corwin DL. Climate change impacts on soil salinity in agricultural areas. Eur J Soil Sci 2021;72(2):842–62.

[102] Teutschbein C, Seibert J. Regional climate models for hydrological impact studies at the catchment scale: a review of recent modeling strategies. Geogr Compass 2010;4(7):834–60.

[103] Salzer MW, Kipfmueller KF. Reconstructed temperature and precipitation on a millennial timescale from tree-rings in the southern Colorado Plateau, USA. Clim Change 2005;70(3):465–87.

[104] Kidson JW. An analysis of New Zealand synoptic types and their use in defining weather regimes. Int J Climatol 2000;20(3):299–316.

[105] Vogel E, Donat MG, Alexander LV, Meinshausen M, Ray DK, Karoly D, et al. The effects of climate extremes on global agricultural yields. Environ Res Lett 2019;14(5):054010.

[106] Priya R, Ramesh D ML. based sustainable precision agriculture: a future generation perspective. Sustain Comput Inform Syst 2020;28:100439.

[107] Sharma A, Jain A, Gupta P, Chowdary V. Machine learning applications for precision agriculture: a comprehensive review. IEEE Access 2020;9:4843–73.

[108] Tassetti AN, Galdelli A, Mancini A, Punzo E, Bolognini L. Underwater mussel culture grounds: precision technologies for management purposes. 2022 IEEE International Workshop on Metrology for the Sea. Learning to Measure Sea Health Parameters (MetroSea) 2022, 3. IEEE; 2022. p. 153–7. Oct.

[109] Pierpaoli E, Carli G, Pignatti E, Canavari M. Drivers of precision agriculture technologies adoption: a literature review. Procedia Technol 2013;8:61–9.

[110] Heim Jr RR. An overview of weather and climate extremes—products and trends. Weather Clim Extremes 2015;10:1–9.

[111] Mandal SK, Maity A. Precision farming for small agricultural farm: Indian scenario. Am J Exp Agric 2013;3(1):200.

[112] Mondal P, Basu M. Adoption of precision agriculture technologies in India and in some developing countries: scope, present status and strategies. Prog Nat Sci 2009; 19(6):659–66.

[113] Foster AD, Rosenzweig MR. Learning by doing and learning from others: human capital and technical change in agriculture. J Polit Econ 1995;103(6):1176–209.

[114] Barnes AP, Soto I, Eory V, Beck B, Balafoutis AT, Sánchez B, et al. Influencing incentives for precision agricultural technologies within European arable farming systems. Environ Sci Pol 2019;93:66–74.

[115] Mills J, Gaskell P, Ingram J, Dwyer J, Reed M, Short C. Engaging farmers in environmental management through a better understanding of behaviour. Agric Human Values 2017;34(2):283–99.

[116] Bongiovanni R, Lowenberg-DeBoer J. Precision agriculture and sustainability. Precis Agric 2004;5(4):359–87.

[117] Arovuori K., Kola J. Policies and measures for multifunctional agriculture: experts' insight. Int Food Agribus Manag Rev 2005;8(1030-2016-82602):21-51.

[118] Prager K, Schuler J, Helming K, Zander P, Ratinger T, Hagedorn K. Soil degradation, farming practices, institutions and policy responses: an analytical framework. Land Degrad Dev 2011;22(1):32–46.

[119] Finger R, Swinton SM, el Benni N, Walter A. Precision farming at the nexus of agricultural production and the environment. Annu Rev Resour Econ 2019;11(1):313–35.

[120] Smith RP. What makes my baby blue. Dartmouth Medicine 2000:26–31.

[121] Klijn F, Kreibich H, de Moel H, Penning-Rowsell E. Adaptive flood risk management planning based on a comprehensive flood risk conceptualisation. Mitig Adapt Strategies Glob Change 2015;20(6):845–64.

[122] Bell VA, Kay AL, Davies HN, Jones RG. An assessment of the possible impacts of climate change on snow and peak river flows across Britain. Clim Change 2016; 136(3):539–53.

[123] Kay AL, Rudd AC, Fry M, Nash G, Allen S. Climate change impacts on peak river flows: combining national-scale hydrological modelling and probabilistic projections. Clim Risk Manag 2021;31:100263.

CHAPTER 13

Precision agriculture technologies: present adoption and future strategies

Muhammad Jehanzeb Masud Cheema[1,3], Tahir Iqbal[2], Andre Daccache[4], Saddam Hussain[3,4,5], Muhammad Awais[6]

[1]*Department of Land and Water Conservation Engineering, Faculty of Agricultural Engineering and Technology, PMAS-Arid Agriculture University Rawalpindi, Rawalpindi, Punjab, Pakistan;* [2]*Faculty of Agricultural Engineering and Technology, PMAS-Arid Agriculture University Rawalpindi, Rawalpindi, Punjab, Pakistan;* [3]*National Center of Industrial Biotechnology, PMAS-Arid Agriculture University Rawalpindi, Rawalpindi, Punjab, Pakistan;* [4]*Department of Biological and Agricultural Engineering, University of California, Davis, CA, United States;* [5]*Department of Irrigation and Drainage, University of Agriculture, Faisalabad, Punjab, Pakistan;* [6]*Research Center of Fluid Machinery Engineering & Technology, Jiangsu University, Zhenjiangv, Jiangsu, China*

13.1 Introduction

The world population is increasing day by day that will be reached 9 billion by 2050 [1]. The world food demand is expected to rise, while the basic input equation of agriculture is changing due to declining agricultural labor availability, waning agriculture productivity, increasing temperature, shrinking weather regimes, land deprivation, and climate change that is challenging every stakeholder especially in developing countries like Pakistan. On the other hand, industries also consume more resources for various applications, thus increasing competition in agricultural production.

Worldwide it is imperative to notice that agricultural land produces 46% of the world food supply while occupying just 18% of cultivated land [2,3]. Global food production has been projected to increase by 70% over the upcoming years to cope with this development. Latest farming methods have not been adopted as these should be, and producers still use conventional strategies based on speculation about their nutritional requirements. Supplying the farms with nutrient inputs is no longer the best option because this leads to heavy use of synthetic fertilizers, high operating costs, unnecessary water use, and environmental degradation [4,5].

In addition, the farms are also facing the problem of lower yields. For instance, the average wheat yield in Pakistan is 1.2 tha^{-1}, Australia has 2 tha^{-1}, 2.4 tha^{-1} (max. 5.3 tha^{-1}) of China, while the United States has 3.1 tha^{-1}. The other cash

Table 13.1 Yield comparison of major crops in Pakistan and China [GoP, knoema.com].

Crops	Pakistan Avg. (max[a]) tha^{-1}	China Avg. (max[a]) tha^{-1}
Wheat	1.26 *(3.1)*	2.4 *(5.3)*
Cotton	0.8 *(1.2)*	1.7 *(3.5)*
Rice	1.1 *(3.5)*	3.1 *(5.1)*
Corn	2.3 *(5.3)*	2.9 *(10.1)*
Sugarcane	28.7 *(74.9)*	34.1 *(n.a.)*
Tomato	4.6 *(13.2)*	13.7 *(37.5)*
Potato	10.9 *(12.3)*	n.a.

[a] The maximum yield is based on the author's observation while farm visits and discussions with the farmers of the two countries. Could be more in other farms that have not been reported.

crops like cotton, rice, corn, sugarcane, and tomato also have similar yield gaps. For example, the maximum rice yield in China is 5.1 tha^{-1}, 3.7 in Brazil, and 5.9 in the United States [6]. A comparison of average and maximum yields of major crops in Pakistan and China is provided in Table 13.1.

One of the reasons for this huge difference in maximum and average yields is efficient use of technologies. Innovative agriculture techniques such as precision agriculture (PA) methods are being adopted worldwide to narrow the yield gaps. PA is a farm management method that uses information technology (IT) to effectively disperse inputs and ensure that soil receive required nutrients at right time for improved health and enhance productivity [7]. China, the United States, and other advanced countries have adopted modern agriculture techniques that revolutionized their agriculture sector. Special focus was given to developing site-specific, precise machinery/technologies termed as precision agriculture technologies (PATs) [8–10].

PA is based on observing, measuring, and responding to inter- and intrafield variability in crops. It aims to define a decision support system (DSS) for whole-farm management to optimize returns on inputs while preserving resources [9]. The three main technical components of PA are (a) data collection through remote sensing (RS) networks, i.e., satellite and unmanned aerial vehicles (UAVs); (b) geographic information system (GIS), where data collection and visualization are carried out using different techniques and software; and (c) modern PA tools such as variable rate applicator that enable site-specific recommendation to be applied. It involves fast, consistent, and circulated measurements to inform farmers and a supplementary overview of their current situation on the cultivated land. The technology frequently used in PA includes integrated electronic communication, wireless sensor networks, precision positioning systems, machine learning (ML) methodology, variable rate technology, UAV, and geographical information systems for well-timed crop management [11].

PA is also being bolstered by new technologies, including internet of things (IoT) and UAVs in crop monitoring, cloud computing, big data, block chain, deep learning, mobile apps, and artificial intelligence (AI). Farm sustainability can be enhanced by integrating traditional agricultural methods with cutting-edge technologies, practices, and economic factors, e.g., UAVs capture aerial photographs of agricultural fields to track crop health, estimate nutrient status, yield, and calculate crop water demand. All of these technologies work together to help make better pre- and postproduction decisions and contribute to the sustainability of farm production system [12–15].

PATs emerged in late 1980s and are now being extensively used from land preparation to harvesting and postharvesting. The United States is the early adopter and technically advanced country using these technologies for the last three to four decades. PAT has also been well adopted in Canada, Australia, and some European countries, including Germany, Sweden, the Netherlands, Denmark, and Finland [16]. However, in developing countries like Pakistan, farmers mostly rely on conventional agriculture technologies, thus resulting in reduced farm yield, increased input cost, and environmental hazards.

In this chapter an effort is made to highlight the concept of PAT and transformation in the agriculture sector from mechanical to PATs. The adoption of these technologies in developed and developing countries will also be discussed. Furthermore, it will also be tried to pen down farmers' preferences for different technologies and their preferred products and the potential future demand for PATs.

13.1.1 Concept of precision agriculture

A number of synonyms are being used to describe PA, including precision farming, smart farming, smart agriculture, climate-smart farming, etc. Now digital agriculture has also emerged that uses *information technology* resources/gadgets to make farming decisions, enabling two-way communication between the decision-makers and machines, thus assisting in doing agriculture, precisely and efficiently, utilizing a combination of sensors system. More precisely, PA is defined by Lee et al. [17] as *"precision agriculture is a holistic, innovative systems approach that assists farmers in managing crop and soil variability to decrease costs, improve yield quality and quantity, and enhance farm income."* Modern technologies, techniques, and economic variables promote agricultural sustainability by combining conventional farming methods with IoT-based machinery. The main advantages of introducing PA toward traditional farming practices are:

- Reduced input cost
- Improved farm workability
- Lowered environmental effects
- Higher profit margins

PATs are very useful for getting maximum yield and profit by adopting proper protocol and procedures. PA is at the center of shaping itself to provide solutions to the overarching problems in agriculture. Growing demand for food and other agricultural products has prompted major advances in agronomic technology and agricultural production, thanks largely to advances in chemistry, genetics, irrigation, and robotics. PA and smart irrigation, in particular, enable farmers to save precious resources without subjecting plants to moisture deficiency [3,18,19].

More recently, introduction of information and communication technology (ICT) and IoT-based services to combine and interpret large amounts of data from many sources makes PA more efficient. They allow farmers and service providers to plan and monitor their entire production system activities more effectively, both operational and economics [20]. These techniques' working process requires a vast knowledge of fertilizer map, soil quality, nutrients, and weeds [21]. Many farmers use nutrient requirement data and compare them with spatial variations of the cultivated area [22]. Such requirements pushed the farmers to move from conservative farming to precision agriculture farming [23]. It hopes to improve soil understanding and crop management by making better-informed decisions [24], thus enhancing crop productivity and profitability.

13.1.2 Developments in agricultural technologies

The evolution of agricultural technology development has taken place from manual sowing to the use of horsepower. The horse, camel, and bull power were used for years for all agricultural operations, i.e., crop sowing, driving the well, cutting fodder, traveling, and moving agricultural goods. During 1960–70s, the agriculture tractors were developed, which revolutionized the agriculture mechanization, enhanced agriculture productivity, and increased crop efficiency (Fig. 13.1).

Klerkx et al. [23] recognized the four core methodologies to agricultural innovation adoption, which include "technology transfer (TT), a technology-oriented approach, characterized the period of agricultural modernization (1950–80s)." The TT imitates the idea of knowledge sharing in a "top-down" type from scientists

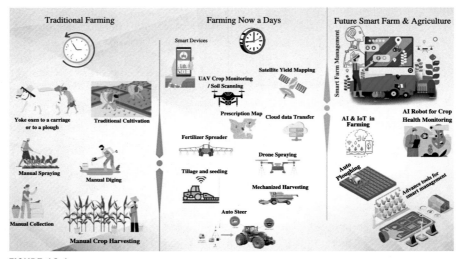

FIGURE 13.1

Evolution of agriculture from traditional to digital.

to the farmers. In this era, the researcher aimed to develop rapid and modern technologies to enhance agricultural production. Moreover, such an attitude was intensely disengaged from the social-politics and institutional context, wherever new pieces of equipment were operational [23,24]. However, after this duration, the scientists examined the system-oriented approach in depth, i.e., agriculture farming system (AFS), agriculture knowledge and information system (AKIS), and agriculture innovation system (AIS) [25].

Using this strategy, farmers were given a new position that shifted their technology usage from a mere user to an adopter [26]. "Top-to-bottom" sharing of new technologies was reproduced in AKIS in the 1990s, where approaches to innovation were not judged as a simple diffusion of knowledge but rather as a mutually beneficial arrangement of skills and expertise across stakeholders. Farmers, manufacturers, and academic institutions that emulated the creative process in and out of agriculture were the primary beneficiaries of this system [27]. The researchers [28] described it statically as an "innovation support infrastructure." In early 1990s, this innovative support infrastructure leads to development of innovations in agriculture technologies, an era of PATs begins.

The precision agricultural technologies have become an important tool to transform agricultural industry and ensure sustainable development of the agricultural supply chain. The development of the global positioning system (GPS) and satellite data helped the farmer for site-specific smart and precise agriculture farming. Nowadays, the term "big data" is widely used in agriculture; right decision is taken with the analysis of the weather, climate, soil, water, crop demand, and market need to grow agriculture crops to meet food security challenges.

PAT may be implemented to more efficiently allocate agricultural production elements, satisfy customer needs individually, and hasten the transition of agricultural production techniques. However, when businesses implement PAT one at a time, various issues with technological tools, structural models, element allocation, talent finances, and supply chain coordination have also emerged. These issues have posed challenges to the advancement of PAT [25] as well as its adoption, especially in developing countries. The farmers have small agricultural landholding and cannot manage the large agriculture machinery in their field for doing precision farming, and Pakistan farm's settings is a typical example [24].

13.2 The transformation from mechanized to precision agriculture technologies

The mechanization and industrialization in the agriculture sector during 20th century led to an increase in the development of large-scale farms, efficacy, productivity, and efficiency [26]. The scarcity of resources and environmental concerns demanded transform alien of conventional model of agriculture farming and its production system. Moreover, the application of newly developed technologies and the role of digitalization led to the transformation from mechanical agriculture to PA [26].

These developed precision agricultural technologies are changing the conventional to a modern farming system that can be effective in ensuring food security. Advances in sensors, wireless communications, and big data analytics are being deployed to agriculture in various ways to help farmers make more informed and accurate decisions. On-farm data and weather/climate conditions can be tracked and analyzed to give farmers guidance on seed selection, soil fertility, crop monitoring, and more accurate spraying of pesticides and fertilizers as well as controlled irrigation commonly termed as "smart agriculture."

The researchers Long et al. [27] built and studied the model for PA to maximize crop production for small- and medium-scale agriculture farming in Indonesia. Moreover, the industrialization and migration of labor force from agriculture farms to industries and cities has resulted in shortage of agriculture labor. This shortage and proven benefits of smart farming encouraged to modernize agriculture and shift from mechanized to PATs.

All the processes in agriculture from land preparation to harvesting have been gradually replaced from tractor-driven implements to more precise IT-enabled devices/machinery that are transforming agriculture to sustainable production system. Various examples of this transformation process from sowing to harvesting are provided in the proceeding sections (Fig. 13.2).

13.2.1 Autosteering technology

Traditionally operation of tractors and other farming machinery were manually operated and depend on the expertise of the operator. However, it was not possible to maintain straight line path while performing farming operations. Sowing in straight lines is not only vital to sow seed at right place but can also be helpful in

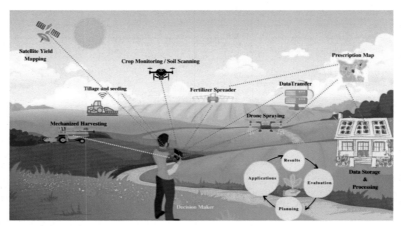

FIGURE 13.2

Conceptual model of farm-level precision agriculture application.

optimizing other farming operations like fertilizers/chemical applications to avoid overlapping and harvesting losses.

With the advent of autosteering technology, all field operations from sowing to harvesting can be done in preciously straight and equidistant lines with centimeter level accuracy. This technology has been widely adopted in various countries and shown fuel and input (water/chemical) saving along with enhancement in crop yield. This technology has resulted in more accurate field works, ease in operation-ability, workability during night and not being affected by bad weather, reduced skips, and overlapping thus reducing inputs like seeds, fertilizer, pesticides, etc. In future the autosteering systems will be a standard feature of all farm tractors. Moreover, testing of driverless autonomous tractors is also being carried out in developed countries especially the United States.

13.2.2 Land leveling to GNSS-based land leveling system

Leveled agriculture fields result in improved efficiency of water and energy resources utilization. In the past, tractor mounted front blade and rear scrappers were used to level the lands. There was no accuracy in cuts and fills and sometime such intervention resulted in soil degradation and lowered yield.

These scrappers were then required with laser emitting device that can scan the field and guide the scrapper to precisely cuts and fills. Such systems are now being replaced by more accurate and precise Global Navigation Satellite System (GNSS)-based leveling systems. The GNSS system uses satellites and ground-based information to provide guidance to the scrapper that provides highly accurate land leveling solution.

13.2.3 Drills to planters and transplanters

Manual sowing (broad casting) was carried out in the past to grow crops. The farmers were unable to grow the crops well in time and therefore lose ideal moisture conditions of the field for sowing, leading to lower crop yield. During 1960s, mechanical drills and planter were developed for row crops like wheat, maize, cotton, etc. In 1990s, bed planters were introduced to grow crops on raised beds with the idea to conserve water, while pneumatic planters have been introduced for precision planting of row crops. Growing of rice was carried out manually in standing water that was labor intensive and time-consuming. The practice has been replaced by direct seeding of rice and more recently with mechanical transplanters.

GPS-guided planters and transplanters helped to plant/transplant crops in straight lines with high precision, thus optimizing seed rate and other farming operations.

13.2.4 Mechanical sprayers to site-specific sprayers

The agrochemicals are extensively used globally for increasing agricultural production and productivity. An important element of crop production is the spraying of insecticides and pesticides during the growing period of the crop to control the insects and weeds, respectively. The weeds grow with the crop and compete with the plant to get nutrients and water for growth, and lower the crop yield, while the insects/pests can destroy the whole crop. Orchards are specifically affected by insects/pest attack.

The existing mechanized spraying systems are being used to spray agrochemicals on the crops and the orchards, yet these systems spray uniformly without addressing the substantial variation in plant population and canopies causing over and under dose. As a result, considerable amount of chemicals drifted with air and deposited into soil and water [28], leading to environmental pollution.

The advent of various sensors (ultrasonic, infrared) and AI/ML has made it possible to develop site-specific, variable rate spraying systems that are smart enough to spray where it is required and as per plant canopy and density. Such sprayers are bit expensive; however, payback period of 2–3 years has made it a better option to replace mechanical sprayers. This variable rate spraying technology is getting popularity among the farming community of the world.

13.2.5 Ground sprayer to aerial spraying system

Ground-based mechanical spraying systems become less effective for spraying in high crops and orchard trees. Top-of-canopy (TOC) spraying especially on the orchards is difficult to achieve due to plant height. Moreover, conventional sprayer boom has uniform spray drift that is less effective for heavy canopies. Conventional application of these agrochemicals therefore results in either over- or underapplication, thus causing environmental problems. The conventional land-spraying machines have been proved to be inefficient for spraying the crops like sugarcane, maize, rice, as well as high-density orchards [29]. Therefore, aerial spraying through UAV is getting popularity worldwide as they can cover entire field without any hindrance [15,29,30].

The UAVs are widely employed for crop monitoring and pesticide/insecticide application in developed countries. In Asian countries, where most fields are small or dispersed, the UAVs have great application potential because their spray rates are 25–50 times lower than traditional sprayers. The aerial spray can contribute to the effectiveness by more than 60% while using 20%–30% less pesticide. In developing countries, such as Pakistani farms, UAV spraying systems are also becoming popular [15,29,30].

13.2.6 Traditional to smart irrigation

In the past, the farmers prefer to use surface irrigation methods, especially flood or boarder, to irrigate most of the grain and fodder crops due to their simplicity and low operational costs. Such surface irrigation methods are less efficient and require regular presence of the irrigator in order to turn off the irrigation once reached at the other end (tail). These less efficient irrigation methods have been replaced with high efficiency irrigation system (HEIS) like drip and sprinkler systems. Both surface and subsurface irrigation systems are in practice in various parts of the world. The center pivot systems are also getting popularity especially for fodder crop in sandy soils.

The advent of sensors and cellular communication has made it possible to revitalize HEIS into more efficient automated system that may reduce irrigator's labor as well as provide irrigation when and how much is required. Sprinklers and drip irrigation systems with sensing devices, nozzles, and controllers that improve coverage and irrigation are part of the smart irrigation system, which monitors moisture-related environment conditions and dynamically adjusts irrigation to optimal levels. To suit the field's water requirements, the intelligent irrigation systems automatically alter watering schedules and run at various times throughout the day. For example, an intelligent irrigation controller will consider site-specific characteristics like soil type, sprinkler rate of application, and other similar considerations while adjusting watering regimens or run periods when temperatures or rainfall drops. These automated systems are used in developed countries like the United States, Israel, China, Europe, etc. The farmers in developing countries are also being attracted to these systems because of their efficacy.

13.2.7 Mechanical harvesters to GPS-based smart harvesting machinery

Shortage of labor, harvesting time, weathering threats, and losses raised another challenge for scientists to develop cutting-edge technology for mass harvesting at the right time to minimize the harvesting losses and timely sowing of next crop. The scientists developed mechanical machines for reaping, threshing, cleaning, and transporting of the crops, i.e., wheat, rice, cotton, maize, etc. Later on, mechanical harvesters, pickers, and shakers were developed to do harvesting/picking quickly.

Mechanical harvesters, pickers, and shakers were adopted significantly due to the increase in harvesting efficiency, clearing large landholding. The shattering and cleaning losses in crops and tree and fruit damage in pickers and shakers were reported up to 25% which was considerably high after huge investment of these machines. Later on, improved versions of combined harvesters were developed considering the crop type, size, and characteristics to minimize harvesting losses.

Incorporation of AI, machine vision, and sensing technology in newly developed harvesters named smart harvester. These smart harvesters not only improve their mechanical efficiency, reduce losses and damages but also give the yield monitoring and grain quality information. Load cells installed on the GPS-guided harvesters provide yield maps to the farmers enabling them to know crop response in terms of yield to inputs provided. These GIS-based yield maps when overlayed on maps of soil properties help to carryout site-specific operations within a particular field.

13.2.8 Remote sensing and GIS/GPS in agriculture

The usage of RS and GIS technologies in agriculture has increased manyfold especially during the last two decades. RS data available through satellites and UAVs equipped with multispectral cameras are helpful to monitor crop growth and identify areas that may be experiencing stress. It helps to identify areas with a high potential for crop production and to plan the expansion of agricultural operations. This information can then be used to target specific areas for further analysis and management.

While, GIS is being used to analyze the spatial relationships between different factors, such as weather patterns, soil types, and irrigation systems, to identify patterns that may be impacting crop growth. It is helpful in identifying the most suitable areas for different crops and to plan crop rotations to improve soil fertility and reduce the risk of pests and diseases. Additionally, GIS can be used to plan and optimize the use of resources, such as water and fertilizer, to maximize crop yields and improve the overall efficiency of agricultural operations.

Both RS and GIS can be used to monitor and manage natural resources such as water, soil, and forests. This includes monitoring water availability, water quality, and water use efficiency. These technologies are being used to monitor and detect pests and diseases, to detect changes in land use and vegetation cover, and to estimate crop yields. Overall, RS and GIS technology can provide farmers with valuable information to make data-driven decisions and improve their agricultural practices. This can lead to more sustainable and profitable agricultural operations and can help to secure food supplies for a growing population.

All mapping, management, and decision-making carried out to do farm-level site-specific agriculture is based on these technologies as PA uses technology and data to optimize crop production. PA uses sensors, such as yield monitors, GPS, and vegetation indices, to collect data on crop growth, soil properties, and weather conditions. These data can then be analyzed using GIS and other tools to create detailed maps of soil properties and crop growth. These maps can be used to identify areas with different soil properties and crop growth patterns and to plan different management strategies for these areas. The farmers can identify areas where these inputs are needed and apply them in a targeted and precise manner. This can help to reduce costs, improve crop yields, and reduce the impact of farming on the environment.

13.3 Adoptability of precision agriculture technologies in the field

The adoption of innovation modern technologies in the fields is one of the crucial factors in the development of agriculture production system [3–5]. It has resulted in economic saving due to optimized inputs, increased yields, and improved environment, ensuring food security and economic sustainability of communities in the developing regions [30]. The farmers are considerably interested in the PA farming techniques at the farm scale to optimize the crop inputs and maximize their profit through developing a proper mechanism for a DSS. As part of PA, RS, geographical imaging, GPSs, data management software, and sensors are all utilized to collect and analyze data.

The farmers who adopted PA technologies in the field (e.g., soil management zoning, autosteer/guidance, variable rate agrochemical/fertilizer application, and yield monitoring) as a whole or in combination have been successful to reap benefits like improved farm management, higher productivity, lower input costs, minimized environmental inputs, as well as improved food quality.

The adoption of the new technologies and innovations depends upon many factors, i.e., farm structure, farmer's characteristics, institutional factors, farmers' education, and interest in modern technology and related information. The main barrier to the wider adaptability of technologies is a conceptual framework for drivers. As testified by various studies, the PATs are widely adopted by well-educated, young, and large farmers who are fully involved and interested in PA farming. The most challenging hurdles are the extraordinary initial investment for purchasing PATs, high learning cost, technical issues with equipment, software, compatibility of equipment with local farm operations, problems or lack of connectivity in rural areas, and user-friendly designs. Another difficulty is that the PA tools need a high level of competencies, knowledge, and skills to handle a large volume of data (big data) collected from PATs.

13.3.1 Adoptability in the developed world

The developed countries have overcome the above barriers and revolutionized their agriculture by adopting PA technologies since 1990s. This has changed the farming culture of agriculture in most of the developed countries. The United States is considered as the country that led the way to implement PATs. The countries like China, Germany, the Netherlands, Australia, Canada, Brazil, etc., are also using advanced agricultural techniques, like GPS, GIS, tractor auto-guided systems, RS, weather and climatic data sets (satellites data), VRTs, variable rate fertilizer, precision irrigation, vertical farming, and hydroponic and aeroponic techniques are being used to maximize the crop yield and net returns.

The United States is leading in PATs as compared to the rest of the world. About 100% of the farmers in the United States have access to high-speed internet, soil sampling, and yield mapping facilities. Among the available PATs, the adoption rate of automatic guidance technology (autosteering) in some states/regions reaches to about 60%–80% recently [31], 74% GPS sprayers, 60% variable rate spraying, and fertigation were offered to dealers, growers, and input providers in 2014–15 [32,33]. According to a survey conducted by Ref. [16], 54% of USA farmers use more than one PA technology on their farms. Fountas et al. [34] reported that about 90% of the yield monitors in the world were operated in the United States. The adoption of soil sampling and yield monitoring maps are widely adopted in the developed world, i.e., the US adoption rate is 98% and 82% (Fig. 13.3).

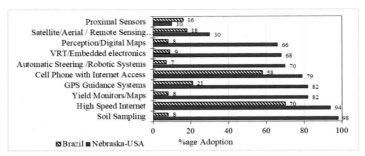

FIGURE 13.3

Adoptability of PA technologies in Brazil comparing to the United States [20].

Data Source: (Brazil: Bolfe ÉL, et al. Precision and digital agriculture: adoption of technologies and perception of Brazilian farmers. Agriculture 2020;10(12):653); (University of Nebraska-Lincoln, USA).

One more common thing, which is in favor of the implementation of PA technologies, is the farm size. The countries such as Australia, Canada, and Europe with bigger farms tended to adopt PA technologies [16]. Australia's 80% of the farmers use autoguidance technology, 98% of Canadian farmers are using GPS guidance, while 36% of Japanese rice farms are protected by UAVs, and around 60% of UK farmers use GNSS-based tractor steering and some other PA gadgets [35,36]. In Western Europe, PAT adoption started a decade after the United States; however, at present about 68% of their small farmers are also using this technology. In Scotland and Ireland, 85% and 43% farmers are adopting the new technologies. While 80% of Danes farmers are growing wheat using precision technologies in Denmark. Adoption rate in Germany is only 30%; however, the German farmers are now motivated to use precision farming technologies and getting quality yields with less inputs.

13.3.2 Adoptability in the developing world

In many advanced countries, modern agriculture is being practiced, but these technologies are currently not widely adopted by farmers across the developing countries [37–39]. The developing countries are still adopting the traditional method of farming, i.e., manual sowing, flood irrigation and walk behind/uniform spraying systems, and traditional crop harvesting. Traditional agriculture farming led toward the loss of inputs and requires a large amount of input material that ultimately pollutes the environment.

The low adoption of precision farming (PF) technologies in developing countries can be explained by a variety of reasons, i.e., a lack of knowledge on the part of the farmers, lack of financial means to make the initial investments, and lack of economic viability of the investments due to small plot sizes or incompatibility of equipment [40].

This is not to say that PATs have not been adopted in emerging nations such as Argentina, Brazil, Turkey, and South Africa. Intelligent irrigation, soil mapping, variable rate irrigation, and yield monitoring technologies were of interest to producers in these nations. In Turkey, only 3% combined harvesters were equipped with yield monitors in 2016, while variable rate applicators were reported as only 20 in number [41]; now these numbers have increased manifold.

Brazil's adoption rate is high as compared to the other developing countries. Brazil growers found much interest in RS procedures, image analysis, and soil mapping. Silva et al. [42] reported that this technology maturity level is 76%, while the adoption rate of VR technologies was to be 29%. Borghi et al. [43] reported that Brazil's yield maps acceptability is around 56%. Albuquerque reported in 2017 that the overall adoption level of PATs in Brazil is about 20%.

Fig. 13.3 shows the clear image of and adoption percentage of different PA techniques in Brazil, a developing country, compared to the United States.

The proximal sensors adoptability in the Brazil was 16%, while 10% in Nebraska, USA. The satellite/RS and digital maps adoptability was 18% and 8% in Brazil, while 30% and 66% in the Nebraska, respectively. However, adoption of VRT, autosteering/robotic system is very high in the United States, i.e., 68% and 70% as compared to the Brazil which is 9% and 7%, respectively. The cell phone with internet access and high-speed internet adoption is higher in both countries (Brazil 58% and 70%; the United States 79% and 94%, respectively).

The PATs are not being adopted by the farmers of developing countries like Pakistan due to the reason that the farmers are less educated, they are not well aware of the operational functions of the new technologies, the modern technologies are more expensive and sensitive for operation, and their maintenance is an issue. The mindset of the farmers is also hindrance to adopt new technologies that need to be changed for widespread adoption of these efficient technologies.

13.4 Case studies from developed and developing countries

The United States is a leading country in the adoption of many innovative technologies. Fountas et al. [44] reported that about 90% of the yield monitors in the world were operated in the United States. The adoption rate of automatic guidance technology in some states/regions reaches to about 60%–80% recently [45,46]. Along with the United States, Australia, Canada, and some European Union countries including Germany, Finland, Denmark, and Sweden have some level of adoption for PATs. Particularly, Leonard [47,48] reported that about 80% of the grain growers use automatic guidance in Australia. Steele [47] indicated that 98% of surveyed farmers used GPS guidance in western Canada.

Zaman et al. [49] developed automated variable rate sprayer in Canada for real-time application of agrochemicals for wild blueberry fields, and they found that the developed VR sprayer was very efficient and precise enough for spot application of the insecticides/pesticides usage in the field. Chattha et al. [50] studied the VR spreader for real-time spot application of granular fertilizer in Canada for wild blueberry and found that fertilizer spreader was efficient and accurate for spot application of fertilizer to increase the farm profitability and reduce the environmental risks. Esau et al. [51] studied the machine vision based smart agrochemical sprayer in Canada for spot application of insecticide/pesticides in wild blueberry fields and revealed that foliar fertilizer and fungicide increased healthy wild blueberry plants by 57.8% and harvestable yield by 137.8%. Fungicide application savings using the smart sprayer for SA were 11.6%. Hussain et al. [52] also designed and developed the SVR sprayer in Canada using deep learning and revealed that 40% spraying liquid can be saved using SVR sprayer.

Wandkar et al. [53] evaluated the newly developed variable rate sprayer in India to check the effect of air velocity on spraying deposition in guava orchard and revealed that air velocity of 35 m/s and hollow cone nozzle was found to deposit higher spray deposition with as compared to other selected air velocities and forward speed of the sprayer. Liu et al. [54] tested the precision VR spraying robot using single 3D LIDAR in the orchard in China and revealed that, the use of VR spraying robot in orchard reduced the insecticide/pesticides application volume and effectively controlled the environmental pollution caused by the uniform spraying of the insecticide/pesticides. Wang et al. [55] used the vision-based VR spraying approach through UAV to apply pesticides in the rice field in China and found that this approach is also applicable through drones and experimental results proved that, VR spraying has reduced the amount of pesticide used during UAV spraying which can provide a reference for PA adoption in future.

Esau et al. [56] utilized the GNSS autosteer to harvest the wild blueberries successfully in Canada. They have compared manual picker with the autosteer and found that autosteer system's accuracy was considerably higher, and its absolute mean pass-to-pass accuracy was 22.7 mm better than manual picker. The implementation of an RTK autosteer guiding system resulted in not just lower yield losses but also higher net profits. Given the peanut prices and production costs considered in this study, a farmer could potentially expect net returns between 94 and 404 USD ha^{-1} higher by using an RTK autosteer guidance system with accuracy within 25 mm for digging peanuts, and net returns between 323 and 695 USD ha^{-1} higher if row deviations of 180 mm are avoided [57].

The observed variations in tractor steering quality based on SF1 and SF2 signals are not significant enough to justify the cost of signal subscription. Although automatic steering using the SF2 signal reduces missing areas, the advantages do not appear to be proportional to the higher costs [58].

13.5 Yield improvement and farm profitability through the adoption of PA technologies

PATs have promising potential to increase farm yield and profitability by applying the right amount of inputs at the right time, improving crop yield with lower production cost (reducing fuel, water, and chemical), and controlling GHG emissions. In traditional farming, the decision is based on conventional practices, regional conditions, and historical data, while PA technologies are well equipped with sensors, cameras, IoT gadgets, GPS, GIS, RS images, and real-time situation of soil and plants. It supports farmers to take the right decisions regarding sowing time, when and how much irrigation to be applied, amount of fertilizer, calibrated dosage of pesticide, and when to harvest resulting in maximum production. The farmers who have adopted PATs (soil sampling, GPS-guided autosteering, real-time tillage and seed monitoring, site-specific crop specific fertilization and spraying, yield monitoring) have succeeded in getting maximum yield and better quality produce [59]. Such products have easy traceability that makes it best suitable export commodity as most of the international trade demands traceability of the agricultural products.

PATs have an overall positive impact on farm profitability. Some of the technologies (automation of section control system, seed metering and distribution, site-specific applications of chemicals) make immediate payback profit to the grower, while some data technologies (soil mapping, RS images, and VR applications) have intangible benefits [18].

Farm profitability can be accessed by understanding and calculating the benefit of cost difference on sowing seed with or without GNSS autosteer, the cost difference between spraying and fertilization with or without VR application, irrigation with or without control system, yield, and quality of products with fewer inputs. Griffin et al. [18] investigated that if a farmer invests in PA technologies with an initial cost of 30,000 USD, subscription cost for software and tools of 500 USD per year, and repair and maintenance costs of 1500 USD per year. He can save up to 11,000 USD per year in seed cost and 500 USD per year in fuel compared to the conventional method. In another example from Pakistan, investment of about 15,000 USD on autosteering technology installed on local tractors can generate additional income of around 8000 USD.

13.6 Problem/issues in adoptability of precision technologies

There are many issues in adopting PATs, and these problems may vary from country to country. The common issues that have been observed in adopting PATs can be summarized as:

- Small agriculture landholdings
- High initial capital cost and operational controls
- Farmer's education as high level of competency is required
- Lack of trained manpower for technical operations of the technology
- The high import price of the technologies and lack of local developers
- Nonavailability of skilled repair services and spare parts
- Lack of connectivity in rural areas
- Most importantly, the conservative mindset and willingness of the end user to adopt new technologies

The adoption rate of technologies associated with PA is lower in the regions having small to medium farm size as these are considered to be the technologies for the large farms. Munz and Schuele [60] estimated that 78 ha farm is economically viable to apply complete PAT package. Furthermore, high costs, lack of digital infrastructure (electricity/internet), awareness, and skill lack in the farmers as well as societal barriers are challenges in scaling PA solutions that require cross-sectoral collaboration among the farmer community, academia, industry, and government. Enabling policies are essential for promoting innovations in agriculture sector and scaling up of technology adoptability. Provision of aggrotech service providers, hands on trainings, and digital capacity building of the farmers is required for large-scale adoption of PATs.

13.7 Future prospects of technology adoption

The farmers of developed countries have understood the importance of PATs and are willing to adopt these technologies. Moreover, technological improvements in these technologies also resulted in higher adoption rate. In the developing countries, it is still a challenge especially due to small landholdings. Most farming units are small-scale farms, and higher profits are important for small-scale farmers [61]. However, now farmers are discussing and keen to use precision technologies especially due to a huge shift of agricultural labor to industries. The lack of skilled labor is compelling the farmers to shift toward using high-mechanized farming. The challenges are mainly operational in nature rather than limitations of technology, itself. Larger farmers, who with the capacity for high initial investment, are shifting toward precision technologies.

In future, joint efforts of government, industry, and farmers are required to be taken in developing countries. Simple, low-cost, scale-free, and user-friendly technologies like management zones delineation, variable rate sprayers, and fertilizer spreaders are required to be introduced. Technology adoption related difficulties can be overcome by providing technical education to the farmers and operators. The academic institutions are best options for skill enhancement in using automated machinery and IoT-based gadgets. Professional vendors and agricultural service providers can be introduced to provide professional support to the farmers. Ecofriendly policy formulation, implementation, and continuous regulations can further enhance future adoption of PAT.

13.8 Conclusions

In this chapter the adoption potential of PATs has been discussed. The PA has the potential to enhance agricultural productivity by around 70% be the year 2050. However, despite of countless financial, social, and environmental benefits, the adoption of PAT is relatively slow in developing countries. The farming community in the developing world is still stuck to conventional farming methods. Due to this conservative mindset, there is a huge gap in agriculture production between developing countries and the developed world. One of the major bottlenecks, i.e., high initial cost of the tools/equipment, could be resolved by providing indigenized solutions and selection of one or more technologies that could be best adoptable in a specific region. It is also observed that various precision technologies are continuously becoming cheaper [61]. These technologies can therefore be introduced stepwise depending upon the farm size and farmers' financial capacity. Establishment of PAT tools is also needed that could be operated through private sector service providers. Enabling policies to promote PATs is required from the government side.

References

[1] Ahirwar S, et al. Application of drone in agriculture. Int J Curr Microbiol Appl Sci 2019;8(1):2500–5.

[2] Döll P, Siebert S. Global modeling of irrigation water requirements. Water Resour Res 2002;38(4):8-1–8-10.

[3] Gebbers R, Adamchuk VI. Precision agriculture and food security. Science 2010;327(5967):828–31.

[4] Abdullah FA, Samah BA. Factors impinging farmers' use of agriculture technology. Asian Soc Sci 2013;9(3):120.

[5] Blandford D, Braden J, Shortle JS. Economics of natural resources and environment in agriculture. In: Agriculture and the environment. Elsevier; 2014. p. 18–34.

[6] Seth A, Ganguly K. Digital technologies transforming Indian agriculture. The Global Innovation Index; 2017. p. 105–11.

[7] Cuong TX, et al. Effects of silicon-based fertilizer on growth, yield and nutrient uptake of rice in tropical zone of Vietnam. Rice Sci 2017;24(5):283–90.

[8] Hedley C. The role of precision agriculture for improved nutrient management on farms. J Sci Food Agric 2015;95(1):12–9.

[9] McBratney A, et al. Future directions of precision agriculture. Precis Agric 2005;6(1):7–23.

[10] Pierce FJ, Nowak P. Aspects of precision agriculture. Adv Agron 1999;67:1–85.

[11] Liaghat S, Balasundram SK. A review: the role of remote sensing in precision agriculture. Am J Agric Biol Sci 2010;5(1):50–5.

[12] Hussain S, et al. Spray uniformity testing of unmanned aerial spraying system for precise agro-chemical applications. Pakistan J Agric Sci 2019;56(4).

[13] Awais M, et al. Comparative evaluation of land surface temperature images from unmanned aerial vehicle and satellite observation for agricultural areas using in situ data. Agriculture 2022;12(2):184.

[14] Awais M, et al. UAV-based remote sensing in plant stress imagine using high-resolution thermal sensor for digital agriculture practices: a meta-review. Int J Environ Sci Technol 2022:1–18.

[15] Li W, et al. A UAV-aided prediction system of soil moisture content relying on thermal infrared remote sensing. Int J Environ Sci Technol 2022:1–14.

[16] Say SM, et al. Adoption of precision agriculture technologies in developed and developing countries. Online J Sci Technol-Jan 2018;8(1):7–15.

[17] Lee C-L, Strong R, Dooley KE. Analyzing precision agriculture adoption across the globe: a systematic review of scholarship from 1999–2020. Sustainability 2021; 13(18):10295.
[18] Griffin TW, Shockley JM, Mark TB. Economics of precision farming. In: Precision agriculture basics; 2018. p. 221–30.
[19] Grisso RD, et al. Precision farming tools: variable-rate application. 2011.
[20] Bolfe ÉL, et al. Precision and digital agriculture: adoption of technologies and perception of Brazilian farmers. Agriculture 2020;10(12):653.
[21] Moran MS, Inoue Y, Barnes E. Opportunities and limitations for image-based remote sensing in precision crop management. Remote Sens Environ 1997;61(3):319–46.
[22] Robertson M, Carberry P, Brennan L. The economic benefits of precision agriculture: case studies from Australian grain farms. Crop Pasture Sci 2007;12:2012.
[23] Vellidis G, et al. A real-time wireless smart sensor array for scheduling irrigation. Comput Electron Agric 2008;61(1):44–50.
[24] Toriyama K. Development of precision agriculture and ICT application thereof to manage spatial variability of crop growth. Soil Sci Plant Nutr 2020;66(6):811–9.
[25] Awais M, et al. Evaluating removal of tar contents in syngas produced from downdraft biomass gasification system. Int J Green Energy 2018;15(12):724–31.
[26] Sunding D, Zilberman D. The agricultural innovation process: research and technology adoption in a changing agricultural sector. Handb Agric Econ 2001;1:207–61.
[27] Long TB, Blok V, Coninx I. Barriers to the adoption and diffusion of technological innovations for climate-smart agriculture in Europe: evidence from the Netherlands, France, Switzerland and Italy. J Clean Prod 2016;112:9–21.
[28] Zhang X, Luo Y, Goh KSJEP. Modeling spray drift and runoff-related inputs of pesticides to receiving water. Environ Pollut 2018;234:48–58.
[29] Sheng H, et al. Loss of labile organic carbon from subsoil due to land-use changes in subtropical China. Soil Biol Biochem 2015;88:148–57.
[30] Thilakarathna MS, Raizada MN. Challenges in using precision agriculture to optimize symbiotic nitrogen fixation in legumes: progress, limitations, and future improvements needed in diagnostic testing. Agronomy 2018;8(5):78.
[31] Miller N, et al. Farmers' adoption path of precision agriculture technology. Adv Anim Biosci 2017;8(2):708–12.
[32] Erickson B, Widmar DA. Precision agricultural services dealership survey results, vol 37. West Lafayette, Indiana, USA: Purdue University; 2015.
[33] Lowenberg-DeBoer J. The precision agriculture revolution. Foreign Aff 2015;94:105.
[34] Fountas S, et al. Farmer experience with precision agriculture in Denmark and the US Eastern Corn Belt. Precis Agric 2005;6:121–41.
[35] Liao M. XAIRCRAFT launched in Japan targeting global precision farming. 2017.
[36] Leonard E. Precision Ag down under. 2014.
[37] Reichardt M, Jürgens CJPA. Adoption and future perspective of precision farming in Germany: results of several surveys among different agricultural target groups. Precis Agric 2009;10:73–94.
[38] Kutter T, et al. The role of communication and co-operation in the adoption of precision farming. Precis Agric 2011;12:2–17.
[39] Loudjani P. Precision agriculture: an opportunity for EU-farmers—potential support with the CAP 2014-2020. 2014.

[40] Takácsné György K, et al. Precision agriculture in Hungary: assessment of perceptions and accounting records of FADN arable farms. Stud Agric Econ 2018;120(1316-2018-2929):47−54.

[41] Akdemır B. Evaluation of precision farming research and applications in Turkey. In: VII international scientific agriculture symposium, "Agrosym 2016," 6−9 October 2016, Jahorina, Bosnia and Herzegovina. Proceedings. University of East Sarajevo, Faculty of Agriculture; 2016.

[42] Silva CB, de Moraes MAFD, Molin JP. Adoption and use of precision agriculture technologies in the sugarcane industry of São Paulo state, Brazil. Precis Agric 2011;12(1):67−81.

[43] Borghi E, et al. Adoption and use of precision agriculture in Brazil: perception of growers and service dealership. 2016.

[44] Fountas S, et al. ICT in Precision Agriculture—diffusion of technology. 2005.

[45] Erickson B, Widmar DAJPUWL. Indiana, USA, Precision agricultural services dealership survey results, vol 37; 2015.

[46] Miller NJ, et al. Adoption of precision agriculture technology bundles on Kansas farms. In: 2017 Annual meeting, February 4−7, 2017, mobile, Alabama. Southern Agricultural Economics Association; 2017.

[47] Steele DJFR. Analysis of precision agriculture adoption & barriers in western Canada, vol 53; 2017.

[48] Abobatta WF. Precision agriculture: a new tool for development. In: Precision agriculture technologies for food security and sustainability. IGI Global; 2021. p. 23−45.

[49] Zaman QU, et al. Development of prototype automated variable rate sprayer for real-time spot-application of agrochemicals in wild blueberry fields. Comput Electron Agric 2011;76(2):175−82.

[50] Chattha HS, et al. Variable rate spreader for real-time spot-application of granular fertilizer in wild blueberry. Comput Electron Agric 2014;100:70−8.

[51] Esau T, et al. Machine vision smart sprayer for spot-application of agrochemical in wild blueberry fields. Precis Agric 2018;19:770−88.

[52] Hussain N, et al. Design and development of a smart variable rate sprayer using deep learning. Remote Sens 2020;12(24):4091.

[53] Wandkar SV, et al. Performance evaluation of newly developed variable rate sprayer for spray deposition in guava orchard. Int J Plant Protect 2017;10(1):96−102.

[54] Liu L, et al. Precision Variable-Rate Spraying Robot by Using Single 3D LIDAR in Orchards. Int J Plant Protect 2022;12(10):2509.

[55] Wang L, et al. Vision-based adaptive variable rate spraying approach for unmanned aerial vehicles. Int J Agric Biol Eng 2019;12(3):18−26.

[56] Esau TJ, et al. Evaluation of autosteer in rough terrain at low ground speed for commercial wild blueberry harvesting. Agronomy 2021;11(2):384.

[57] Ortiz BV, et al. Evaluation of agronomic and economic benefits of using RTK-GPS-based auto-steer guidance systems for peanut digging operations. Precis Agric 2013;14:357−75.

[58] Lipiński AJ, et al. Precision of tractor operations with soil cultivation implements using manual and automatic steering modes. Biosyst Eng 2016;145:22−8.

[59] Beluhova-Uzunova R, Dunchev D. Precision technologies in soft fruit production. Sci Pap Ser Manage Econ Eng Agric Rural Dev 2020;20(3):131−7.

[60] Munz J, Schuele H. Influencing the success of precision farming technology adoption—a model-based investigation of economic success factors in small-scale agriculture. Agriculture 2022;12(11):1773.

[61] Mizik T. How can precision farming work on a small scale? A systematic literature review. Precis Agric 2022:1–23.

Glossary

Activation: A mathematical function which is used to introduce a nonlinearity into a neural network, allowing it to model complex patterns.

Algorithm: A set of instructions or steps for solving a problem or achieving a goal.

Ambient Air Temperature: Ambient temperature is the air temperature of any object or environment where equipment is stored.

Approximation: A value or quantity that is nearly but not exactly correct.

Artificial Intelligence: The theory and development of computer systems able to perform tasks that normally require human intelligence, such as visual perception, speech recognition, and decision-making.

Artificial Neural Network: A type of algorithm used in machine learning that is inspired by the structure and function of the brain. It consists of many interconnected "neurons" that process and transmit information, just like the neurons in the human brain.

Autoguidance: A technology that uses GPS and other sensors to automatically guide agricultural machinery such as tractors in the field.

Backpropagation: An algorithm that is used to adjust the weights of the connections between the nodes in a neural network in order to minimize the error between the predicted output of the network and the actual output.

Classification: A machine learning approach in which a model is trained to assign a class label to examples from a predefined set of classes.

Climate Change: Climate change refers to long-term shifts in temperatures and weather patterns. These shifts may be natural, such as through variations in the solar cycle.

Commercialization: The process of managing or running something principally for financial gain.

Convolutional Neural Network: A type of artificial neural network that is particularly well-suited for image recognition and processing tasks.

Cultivation: Cultivation is the act of growing something or improving its growth, especially crops.

Decision Tree: A type of machine learning algorithm which uses a flowchart-like tree structure to make decisions based on the features of the input data.

Deep Learning: A branch of machine learning that makes use of multilayered artificial neural networks to learn from a lot of data. These algorithms are created to identify patterns in data and generate predictions based on that information. They are inspired by the structure and operation of the brain.

Deep Learning Framework: Deep learning (DL) frameworks offer building blocks for designing, training, and validating deep neural networks through a high-level programming interface.

DGPS: An advanced form of GPS navigation which provides greater positioning accuracy than the standard GPS. DGPS relies on error correction transmitted from a GPS receiver placed at a known location.

Discriminant Analysis: A statistical technique that is used to classify observations into one of two or more categories, based on the values of one or more predictor variables.

Disease: An abnormal condition that affects the health and development of a plant, and that is caused by a pathogen or other agent.

Drip/trickle irrigation: An irrigation system in which water is applied directly to the root zone of plants by means of small emitters in the form of droplets.

DSSAT: Decision Support System for Agrotechnology Transfer (DSSAT) software which is used to simulate the crop dynamics under different climatic and irrigation conditions.

Ensemble Methods: Machine learning techniques that combine the predictions of multiple models to make more accurate predictions than any individual model could.

Evapotranspiration: Evapotranspiration includes water evaporation into the atmosphere from the soil surface, evaporation from the capillary fringe of the groundwater table, and evaporation from water bodies on land.

Exclusive OR: A logical operation that outputs true only when the inputs differ (one is true, the other is false).

Feature Importance: Feature Importance refers to techniques that calculate a score for all the input features for a given model—the scores simply represent the "importance" of each feature.

Feature Space: A set of all possible feature combinations that can be used to describe the data.

Feedback Loop: A system in which the output of a process is used as input to influence the same process.

Food Security: Food security states to the availability of food in a country and the ability of individuals within that country to access, afford, and source adequate foodstuffs.

Fruit Losses: Loss of fruits and vegetables as "that weight of wholesome edible product (exclusive of moisture content) that is normally consumed by human.

Fuzzy C-Means: An extension of the k-means clustering algorithm that allows for "fuzzy" assignment of data points to clusters, rather than strict assignments as in k-means.

Gaussian Mixture Models: A statistical model that represents a distribution as a mixture of multiple normal ("Gaussian") distributions.

Geographical Information System: A geographic information system is a computer-based software that uses feature characteristics and position data to develop maps.

Geospatial Technology (GST): GST refers to all emerging technologies such as Remote Sensing (RS), Geographical Information Systems (GIS), and Global Positioning Systems (GPS), that help the user in the collection, analysis, and interpretation of spatial and temporal data.

Global Positioning System: A GPS is the only system today able to show our exact position on the Earth anytime, in any weather and anywhere.

Gradient Boosting: Gradient boosting is a type of machine learning boosting. It relies on the intuition that the best possible next model, when combined with previous models, minimizes the overall prediction error.

Groundwater: Water held underground in the soil or in pores and crevices in rock.

Hand Raking: A small rake with a short handle, hand rakes typically have a metal head and a wooden handle or fiberglass handle to hold harvested crop.

Harvesting: Gathering a mature crop from the fields is harvesting.

Herbicide: Chemicals that are used to kill or control the growth of plants, especially weeds.

Hydra probe II soil sensor: Used to measure soil characteristics (temperature, moisture, salinity level, and conductivity).

Hyperparameter: Hyperparameters are parameters whose values control the learning process and determine the values of model parameters that a learning algorithm ends up learning.

Hyperplane: A subspace of one dimension less than the feature space which divides the feature space into two halves.

ImageNet: A large dataset of images used for training and evaluating computer vision models, particularly convolutional neural networks.

Infrared: Electromagnetic energy with wavelengths longer than those of visible light, but shorter than those of radio waves.

Integrated pest management (IPM): Integrated pest management, or IPM, is a process you can use to solve pest problems while minimizing risks to people and the environment. IPM can be used to manage all kinds of pests anywhere—in urban, agricultural, and wildland or natural areas.

IoT: Internet of Things (IoT) describes the digital connection of physical objects (things) with sensors and software embedded with internet to connect and exchange data with other devices.

Irrigation: Artificial application of water to land or crops for the purpose of supporting plant growth and increasing crop yields.

K-Means Clustering: A method of unsupervised machine learning that is used to partition a dataset into a specified number (k) of clusters.

K-Nearest Neighbors: A type of machine learning algorithm which stores all available cases and classifies new cases based on a similarity measure.

Livestock: Domesticated animals that are raised for their economic value, such as for their meat, milk, eggs, or other products.

Machine Learning: A form of artificial intelligence in which data is analysed and learned by using algorithms rather than being explicitly programmed.

Management Practices: Management practices usually refers to the working methods and innovations that managers use to improve the effectiveness of work systems.

Management Zones: The management zones are defined by dividing the field into different parts to manage spatial variability within the fields.

Mechanical Harvesting: Mechanical harvesting systems are designed to achieve the mass removal of the commodity during the harvesting season at once. This method has been practiced by shaking the trunks, limbs, and canopies of plants.

Mechanization: The introduction of machines or automatic devices into a process, activity, or place is known as mechanization.

Meteorological: Relating to the branch of science concerned with the processes and phenomena of the atmosphere, especially as a means of forecasting the weather.

Microirrigation systems: These systems are referred as High Efficiency Irrigation Systems (HEISs) that include drip, surface/subsurface, and sprinkler irrigation methods.

Mitigation: Mitigation means reducing the risk of loss from the occurrence of any undesirable event.

Monocrop: A type of agricultural system in which a single crop is grown over a large area, rather than a diversity of crops being grown in the same space.

Multilayer Perceptron: Multilayer perceptron (MLP) is a supplement of feed forward neural network. It consists of three types of layers—the input layer, output layer, and hidden layer to map functions between input and outputs.

Neural Network: see "Artificial Neural Network."

Neuron: A basic unit that performs a simple computation on input data and produces an output, the building blocks of an artificial neural network.

Perceptron: A type of algorithm that can be used for pattern recognition and classification tasks. It works by learning a set of weights that can be used to predict the class of a given input based on its features.

Precision Agriculture: Precision agriculture is a method of farming that uses technological innovations including GPS guidance, drones, sensors, soil sampling, and precision machinery to grow crops more efficiently.

Precision Irrigation (PI): PI is the application of need-based and precisely calculated amount of water (or nutrients) to a plant (or irrigation zone) at an appropriate time and location.

Prescription Map: A document that outlines the management practices that should be followed on a particular piece of land. Prescription maps are often used in agriculture and forestry to guide the use of fertilizers, pesticides, and other inputs, as well as to outline specific management practices such as irrigation, planting, and harvest.

Random Forest: A set of decision trees trained on a randomly selected subset of the total training data.

Rectified Linear Unit: A type of activation function commonly used in neural networks.

Recurrent Neural Network: Recurrent neural networks recognize data's sequential characteristics and use patterns to predict the next likely scenario.

Regression: A machine learning approach that is used to predict a continuous numerical value.

Reinforcement Learning: A type of machine learning in which an agent learns to interact with its environment in order to maximize a reward signal.

Relative Humidity: The amount of water vapor present in air expressed as a percentage of the amount needed for saturation at the same temperature.

Remote sensing: Remote sensing is used to collect field data without physically contacting the target, i.e., soil or plant.

Runoff: Something that drains or flows off, as rain that flows off from the land in streams.

Scouting: Scouting is the process of routinely checking crops for pests and disease to inform management decisions.

Sensor: A piece of equipment that monitors and reacts to a physical characteristic or phenomenon, such as temperature, pressure, motion, or light.

Spatial Variability (SV): SV occurs when a quantity that is measured at different spatial locations exhibits values that differ across the locations.

Sprinkler Irrigation: Sprinkler irrigation is an overhead irrigation system that sprays water on the land or crop in a manner similar to rain.

Supervised Learning: A machine learning approach in which a model is trained to make predictions based on labeled examples.

Support Vector Machine: A type of machine learning algorithm which separates data into classes by fitting a hyperplane which maximally separates the classes.

Sustainability: Sustainability means meeting our own needs without compromising the ability of future generations to meet their own needs.

Topographical Aspects: In physical geography, aspect is the compass direction that a topographic slope faces, usually measured in degrees from north. Aspect can be generated from continuous elevation surfaces.

Topography: The arrangement of the natural and artificial physical features of an area.

Unsupervised Learning: A machine learning approach in which a model is trained to discover patterns or relationships in a dataset without the use of labeled examples.

Variable-Rate Technology (VRT): Variable-rate technology is defined as the application of a suitable number of fertilizers at the right time in a precise manner, which will reduce inputs, costs, and unfavorable environmental impacts and improve crop yield and quality.

Vegetative Growth: Vegetative growth is the growth of leaves, stems, and roots.

Water-Use Efficiency (WUE): WUE is crop yield per unit of water use also named as water productivity (WP).

Weed: A plant that is growing where it is not wanted.

Yield Monitoring: Yield Monitoring is an aspect of precision agriculture that helps to provide farmers with adequate information to make educated decisions about their fields.

Index

Note: Page numbers followed by "*f*" indicate figures and "*t*" indicate tables.

A

Adaptability
  developed world, 243–246
  developing world, 244–246
  farm probability, 246–247
  green technologies, 243
  problem/issue, 245–246
  yield improvement, 246
Aerial image analysis, 37
Agricultural practices and products, 2
Agricultural resources management
  crop variability, 4–5
  farmlands, 4
  field agricultural activities, 4
  nutrient management, 5
  soil management, 5
  supplemental light and CO_2, 6–7
  water management, 5–6
Agricultural revolution, 1
Agricultural sustainability, 2
  agricultural enterprises, 190
  farm household incomes, 204
  financial and commercial sustainability, 196–204
  technology adoption, 190–191
    commercial barriers, 198
    economic dynamics, 204–205
    factors influencing, 194–196
    Pakistan, 191–193
    Punjab, 191–193
    soil and nutrient smart, 193
    water-smart, 193
Agriculture knowledge and information system (AKIS), 236
Agrochemicals, 7
Airborne and space-borne sensors, 73
Ant colony optimization-online sequential extreme learning machines (ACO-OSELM), 113
Artificial intelligence (AI), 11, 30, 65–66, 85–86, 105. *See also* Machine learning (ML)
  unmanned aerial vehicles (UAVs), 65–66
  and variable rate technologies (VRTs), 112–113, 116
Artificial neural networks (ANNs), 144–145
Auto steering system, 77
Autoguidance, 151–152
Automated harvesting system, 127–128
Automatic irrigation scheduling techniques, 85–86
Autosteer systems, 213–214

B

Basin irrigation, 88, 90f
Best crop management practices, 127–128
Big data analytics, 75–76
Border irrigation, 88, 90f
Bubbler irrigation, 89

C

Capital budgeting techniques, 197–198
Carbon footprints, 2
Chlorophyll index, 79f
Climate change
  impacts, 219–221
  threats, 3
Computer-based forecasting model, 40
Conventional agriculture, 7
Conventional fertilizer, 218
Convolutional neural networks (CNNs), 112, 145, 174–175
Correlation-based feature importance methods, 175–176
Crop economy, 30
Crop stress monitoring system, 78
Crop yield comparison, Pakistan and China, 232t
Cropping systems, 105–107
Crops yield, 55

D

Decision support system (DSS), 187, 231–232
Decision trees, 142–143, 143f
Deep learning algorithms, 50, 75–76
Deep learning frameworks
  Keras, 172
  MatLab, 173
  PyTorch, 172
  TensorFlow, 172
Deep neural network approaches
  artificial neural networks (ANNs), 144–145
  convolutional neural networks (CNNs), 145
Differential global positioning system, 8
Digital technologies, 80
Discriminant Analysis (DA) model, 146–147
Disease-water stress index (DSWI), 46

255

Drip irrigation, 89, 93f
Drone technology, 116
Drones-based granular fertilizers, 104

E

Ecological impacts
    conventional agriculture practices, 198−201
    precision agriculture, 201−203
Economic dynamics, 204−205
Economic threshold levels (ETLs), 44
Ensemble methods, 143
Environmental benefits, 3
Environmental impacts
    air, 218
    climate change impacts, 219−221
    conventional agriculture practices, 198
    groundwater, 215−217
    precision agriculture, 201
    surface water, 214−215
Environmental pollution, 3
Evapotranspiration (ETo) modeling, 180−181
Exchangeable Image File Format (EXIF) metadata, 39−40
Extreme Learning Machine (ELM), 148−149

F

FAO food insecurity experience scale, 3, 4f
Farm sustainability, 232
Farming practices, 1−2
Fertilizer and pesticide applications
    herbicide and insecticide application, 203
    irrigation, 202
    soil erosion, 202
Financial and commercial sustainability
    capital budgeting techniques, 197−198
    cropwise yield gains and cost savings, 196−197
    environmental impacts, 198
Fixed-wing multirotor hybrid unmanned aerial vehicles, 57
Fixed-wing unmanned aerial vehicles (UAVs), 56, 57f
Flood irrigation, 88
Food security challenges, 3
Furrow irrigation, 88, 90f
Fuzzy C-Means clustering, 144

G

Gaussian Mixture Models (GMMs), 144
Geographic information systems (GIS), 10, 26, 45, 73, 241−242
Geospatial technologies (GST)
    pest and disease management
        case studies, 40
        data and information sharing, 51
        pests and disease monitoring, 45−47
        plant disease identification, 50
        sensing systems, 45−47
    spatially variable rate technology, 48
    unmanned aerial vehicles (UAVs), 49−50
Geostatistical methods, 27−28
Geostatistics, 20
Global Navigation Satellite System (GNSS), 37, 73
Global positioning system (GPS), 73, 211−212
Gradient boosting, 175
Gravity irrigation, 88
Green revolution, 1, 71
Greenhouse gases (GHG), 31
Ground robots, 153−154
Groundwater
    depletion, 215−216
    nitrate concentration, 217t
    nitrate concentrations, 216
    nitrogen fertilizers, 215−216
    VRT studies, 216−217
Groundwater level (GWL) modeling, 178−180, 182−183

H

High Efficiency Irrigation Systems (HEISs), 88
High-performing irrigation system, 88
Hybrid fixed-wing multirotor unmanned aerial vehicles, 57
Hyper-spectral sensing, 39

I

Image-based approach, 47
Information and communication technologies (ICTs), 23, 75
Information technology resources/gadgets, 233−234
Infrared sensor system, 108−109
Integrated pest management (IPM), 43−45, 213
Intercatchment wastewater transfer (ICWT), 181
Irrigation, 201. *See also* Precision irrigation (PI)
    methods, 88
    micro-irrigation systems, 88, 91−93
    pressurized irrigation method, 88−91
    surface irrigation methods, 87, 89
    water-use efficiency (WUE), 86
Irrigation management, 29−30

K

Keras, 172
k-Nearest Neighbors (kNN) algorithm, 143
Knowledge extraction methods, 175–176

L

Land surface temperature (LST), 47
Leaf hopper index (LHI), 46
Least squares support vector machine (LS-SVM) method, 112
LIDAR sensors, 108, 109f
Light environment system, 7
Linear regression, 142
Livestock management, 9–10, 147–148
Long-short-term memory (LSTM) neural networks, 169–170

M

Machine learning (ML), 30–31
  agriculture research, 146t
  architectures, 173–175
  classification algorithms, 171
  convolutional neural network (CNN), 174–175
  Discriminant Analysis (DA) model, 146–147
  ethical concerns and challenges, 182
  gradient boosting, 175
  groundwater estimation, 182–183
  hydrological fields
    evapotranspiration (ETo) modeling, 180–181
    groundwater level (GWL) modeling, 178–180, 182–183
    rainfall-runoff modeling, 178
    water resource management, 181
  livestock management, 147–148
  model development
    data preprocessing, 175
    data split and model development c, 176
    feature importance, 175–176
    hyperparameter tuning, 177
  multilayer perceptron (MLP), 169–170, 173
  recurrent neural networks (RNNs), 174
  regression analysis, 171
  soil management, 148
  supervised machine learning, 142–143
  technological developments, 170
  unsupervised machine learning, 144
  water management, 148–149
Machine vision smart sprayer, 109
Management philosophy, 72
Management zones (MZs), 27–28
Map-based variable rate technologies, 238
MatLab, 173
Mean squared error (MSE), 142
Mechanical harvesting, 237
Mechanical planting and transplanting, 236–237
Mechanical sprayers, 237
Micro-irrigation systems, 88, 91–93
Millennium Development Goals (MDGs), 2
Modified soil-adjusted vegetation index (MSAVI), 77, 78f
Moisture-based smart irrigation, 240–243
Multilayer perceptron (MLP), 173
Multirotor unmanned aerial vehicles (UAVs), 56
Multivariate technique, 27–28

N

Nitrate leaching, 8
Nitrogen cycle and pathways, 212, 212f
Nitrogen fertilizers, 215–216
Normalized Difference Vegetation Index (NDVI), 26, 99, 152–153
Nutrient management, 5, 221

O

Online Sequential Extreme Learning Machine (OSELM) model, 114f
Optical sensing techniques, 41
Optimized Soil-Adjusted Vegetation Index (OSAVI), 46
Organic fertilizer, 218

P

Perceptrons, 143
Pest management
  optical sensing techniques, 41
  optical sensors, 42
  plant genotyping and phenotyping, 41–42
Pestlytics, 44
Plant water stress, 75
Plant-based data sensors, 97–98
Precise seeding system, 77
Precision agriculture evolution, 188–189
Precision irrigation (PI), 154–155
  assessment, 96
  benefits, 94
  components, 94
  control, 95
  cycle of, 95f
  data acquisition tools
    plant-based data sensors, 97–98
    remotely sensed data, 99
    soil moisture detections, 98
    weather data acquisition, 97
  data collection, 94–95

Precision irrigation (PI) (*Continued*)
  data interpretation/analysis, 95
  data requirement challenges, 99
  definition, 93–94
  tools and technologies, 96
Prescription maps, 28
Pressurized irrigation, 88–91
Priori knowledge-based methods, 175–176
Proximal sensing, 13–14
Proximal vigor sensors, 106
PyTorch, 172

Q

Quadcoptor unmanned aerial vehicles (UAVs), 56f

R

Rainfall-runoff modeling, 178
Real-Time Kinematic assisted GPS (RTK-GPS), 21
Real-time kinematic (RTK)–based GPS control, 37–38
Recurrent neural networks (RNNs), 169–170, 174
Red-edge chlorophyll index, 79f
Reinforcement learning, 171
Remote sensing technique, 39, 48, 55, 73–75, 241–242

S

Scouting techniques
  insect/pest assessment and control, 44
  pest identification, 43–44
Seeding technology, 12–13
Sensor-based variable rate technologies, 238–239
Single-rotor unmanned aerial vehicles (UAVs), 56
Smart irrigation
  moisture-based smart irrigation, 240–243
  weather-based smart irrigation, 240
Smart systems, 72
Social and gender dynamics, 205–206
Soil and crop variability
  data analysis, 27–28
  factors affecting, 21
  fertilizer application, 20–21
  information and communications technologies, 23
  management zones, 27–28
  monitoring, 21
  precision agriculture technology
    climate change, 31–32
    data collection, 25
    farming tools, 25–26
    geographic information system (GIS), 26
  irrigation management, 29–30
  remote sensing, 26
  soil sampling and analysis, 25
  variable-rate application technology, 28
  traceability system, 23
Soil erosion, 202
Soil management, 1, 5, 76–77, 148
Soil moisture detections, 98
Soil spatial variability. *See* Soil and crop variability
Spatially variable rate technology, 48
Spot specific agrochemical application, 155–158
Sprinkler irrigation, 89–91
State-of-the-art conceptual system
  auto steering system, 77
  crop stress monitoring system, 78
  precise seeding system, 77
  soil management system, 76–77
  variable rate fertilizer application, 77
  variable rate irrigation, 77
  variable rate sprayer, 77
Subsurface drip irrigation (SDI), 93
Subsurface leaching, 8
Subsurface water contamination studies, blueberry cropping system, 106
Supervised machine learning, 142–143
Supplemental lighting system, 7
Support vector machines (SVMs), 142, 142f
Support vector regression (SVR), 13–14
Surface irrigation methods, 87, 89
Sustainable agriculture, 32
Sustainable Development Goals (SDGs), 2

T

Tasseled Cap Transformation (TCT)–based measures, 47
Technologies development
  agriculture development, 236
  digital technologies, 234f, 235
  system-oriented approach, 236
  technology transfer (TT), 236
Technology transfer (TT), 236
TensorFlow, 172
Trickle irrigation, 89, 91–92

U

Ultrasonic sensory system, 107–108, 108f
Unmanned aerial systems (UASs), 238–239
Unmanned aerial vehicles (UAVs), 13, 49–50, 152–153
  artificial intelligence (AI), 65–66
  challenges and limitations, 66

chemical spraying application, 63–64
crop health monitoring, 58–60
disease detection, 63
fixed-wing, 56, 57f
fixed-wing multirotor hybrid, 57
hybrid fixed-wing multirotor, 57
microtechnology advancements, 58
multirotor, 56
pest detection, 62–63
single-rotor, 56
for spraying, 237
types, 56–57
yield estimation, 60–61
Unsupervised machine learning, 144

V

Variable rate agrochemical applications, 155–158
Variable rate application (VRA) technology, 37–38
Variable rate fertilizer application, 77
Variable rate irrigation, 77
Variable rate sprayer, 77
Variable rate technologies (VRTs), 12, 73, 237–238
    adaption of, 110–113, 111t
    and artificial intelligence, 112–113
    artificial intelligence (AI), 105
    biotic and biotic stress monitoring, 104
    crop management actions, 103–104
    cropping systems, 105–107
    development, 105–109
    drones and copter systems, 104
    farmers and researchers, 113–115
    map-based, 238
    remote sensing and spatial variability, 104
    seed density and nitrogen applications, 104
    sensing technologies, 107–109
    sensor-based, 238–239
    site-specific and subfield scale applications, 104
    small land holdings, 116
    spatial variability
        machine learning, 30–31
        management zones (MZs), 28
        prescription maps, 28
        process control technologies, 28
        sensor-based method, 28
        variable-rate, 28
    user-friendliness, 111–112
    variable rate (VR) applicators, 103–104
Vegetation indices (VIs), 46

W

Water footprints, 2
Water index (WI), 46
Water management, 5–6, 148–149
Water-smart, 193
Water-use efficiency (WUE), 86
Weather-based smart irrigation, 240
Weed-IT technology, 48
Wild blueberry
    cultivation and harvesting, 124–125
    fruit losses
        picking efficiency, 134–135
        yield loss prediction, 135–136
        yield mapping and mitigation, 136–137
    hand-raking, 124–125, 125f, 128
    mechanical harvesters
        factors affecting, 130–134
        field topography and vegetative conditions, 134
        history, 128–129
        human-induced factors, 133
        mechanical factors, 134
        temperature and relative humidity, 131
        weather-related factors, 131–132
        working principle, 129–130
    mechanical harvesting, 123, 125f
    ripeness detection, deep neural networks
        convolutional neural networks (CNNs), 149–150
        fruit production, 149
        harvest timing, 149
        You Only Look Once (YOLO) networks, 150
    ripening
        factors affecting, 126–127
        firmness, 126–127
        maturity indices, 126
        meteorological factors, 126–127
World food demand, 231

Y

Yield monitoring system, 78

Printed in the United States
by Baker & Taylor Publisher Services